▶ **Sustainability and Well-Being**

DOI: 10.1057/9781137308993

Other Palgrave Pivot titles

DOI: 10.1057/9781137308993

palgrave▶pivot

Sustainability and Well-Being: The Middle Path to Environment, Society, and the Economy

Asoka Bandarage

palgrave
macmillan

DOI: 10.1057/9781137308993

First published 2013 by
PALGRAVE MACMILLAN

Palgrave Macmillan in the UK is an imprint of Macmillan Publishers Limited, registered in England, company number 785998, of Houndmills, Basingstoke, Hampshire RG21 6XS.

Palgrave Macmillan in the US is a division of St Martin's Press LLC, 175 Fifth Avenue, New York, NY 10010.

Palgrave Macmillan is the global academic imprint of the above companies and has companies and representatives throughout the world.

Palgrave® and Macmillan® are registered trademarks in the United States, the United Kingdom, Europe and other countries

ISBN: 978–1–137–30900–6 EPUB
ISBN: 978–1–137–30899–3 PDF
ISBN: 978–1–137–30898–6 Hardback

This book is printed on paper suitable for recycling and made from fully managed and sustained forest sources. Logging, pulping and manufacturing processes are expected to conform to the environmental regulations of the country of origin.

A catalogue record for this book is available from the British Library.

A catalog record for this book is available from the Library of Congress.

www.palgrave.com/pivot

DOI: 10.1057/9781137308993

Dedicated to Past, Present and Future Generations

DOI: 10.1057/9781137308993

We stand at a critical moment in Earth's history, a time when humanity must choose its future. As the world becomes increasingly interdependent and fragile, the future at once holds great peril and great promise. To move forward, we must recognize that in the midst of a magnificent diversity of cultures and life forms we are one human family and one Earth community with a common destiny. We must join together to bring forth a sustainable global society founded on respect for nature, universal human rights, economic justice, and a culture of peace. Towards this end, it is imperative that we, the peoples of Earth, declare our responsibility to one another, to the greater community of life, and to future generations.

Preamble, THE EARTH CHARTER (2000)

DOI: 10.1057/9781137308993

Contents

DOI: 10.1057/9781137308993

List of Figures

▶

DOI: 10.1057/9781137308993

Acknowledgments

I am grateful to all the individuals and circumstances in my life that enabled me to write this book. While I cannot individually thank all the colleagues, student assistants, friends, and relatives who helped, I would like to acknowledge at least some of them here: my friend and colleague Natalie Pere for encouragement and support throughout the various stages of the manuscript; Christina Brian, publisher, and Amanda McGrath at Palgrave Macmillan for patience and continuing support; Vidhya Jayaprakash and team for the production of the book; anonymous readers of the manuscript for helpful comments and suggestions for improvement; Peter Whitten for valuable editorial support; Donna Sholk and Naomi Taub for research assistance; Kevin Cedeno of the Computer Complex at American University for help with graphics; participants of the faculty seminar of the Center for Health, Society and Risk at American University for feedback on a presentation based on this work; the monks of Washington Buddhist Vihara and Wheaton International Buddhist Centre for discussions of the Middle Path; Mahes and Neville Ladduwahetty, Donald Jennings, Sajed Kamal, Filomina Steady, Roberto Savio, and Ven. Bhikku Bodhi for comments and/or endorsements; James Babson, Rona Fields, John Haywood, Tuyet Chen, and James Swanson for their support; all my teachers, especially the great meditation teacher S. N. Goenka; yoga teacher Neva Ingalls; my late father; my mother and siblings; and my son Maithri, for putting up with me and my preoccupation.

Preamble to the Earth Charter 2000 reprinted with permission from Earth Charter International www.earthcharter.org.

Materials from Asoka Bandarage, *Women, Population and Global Crisis: A Political-Economic Analysis* used with permission from the publisher, Zed Books.

Figure 1 (a) and (b) adapted and reprinted from Bob Giddings, Bill Hopwood, Geoff O'Brien, 'Environment, Economy and Society: Fitting Them together into Sustainable Development', *Sustainable Development*, vol. 10, no. 4, 2002, pp. 187–196 with permission from publisher, John Wiley and Sons.

DOI: 10.1057/9781137308993

1

Introduction: Environment, Society, and the Economy

Abstract: *This chapter refers to the depth of the current crisis of environmental, social, and economic collapse; limitations of current approaches; and the need for alternative cultural and philosophical perspectives. It defines the concepts of sustainability and human well-being, their inseparability, and the need for an integrated approach to environment, society and the economy. It questions the dominance of the economy over the environment and society emphasizing the need for a transformation of the growth-driven economic system toward more balanced alternatives. It calls for an ethical reorientation to build a partnership-based society that upholds sustainability and well-being. It points to the "Middle Path" drawn from the Buddhist teachings for guidance in that direction.*

Bandarage, Asoka. *Sustainability and Well-Being: The Middle Path to Environment, Society, and the Economy.* Basingstoke and New York: Palgrave Macmillan, 2013. DOI: 10.1057/9781137308993.

Humanity has achieved incredible technological and material growth. Yet the ecosystem and human communities are collapsing due largely to that very advancement, causing insecurity, fear, and conflict across the world. As human beings become more and more the appendages of technology and the global market, we face an existential crisis of what it means to be human in nature. The challenge we face is not the further acceleration of competitive, economic, and technological growth and the creation of a postnature, "post-human" world[1] but a fundamental transformation to a balanced and ethical path of social and psychological development.

But as we face that challenge, we live in a time of intellectual and ideological myopia and confusion. Narrowly based social movements present themselves in the guise of upholding truth and freedom while actually fomenting fear, inciting divisions, and fostering ethno-religious and nationalist conflict. Analyses that go to the root of our predicament have virtually disappeared, even in academia.

The unprecedented challenge facing the world—namely, the transformation of the global economy to a sustainable basis[2]—is not receiving the attention that it urgently requires. Conventional academic and policy analyses focus on separate aspects of the global crisis rather than approaching it comprehensively as a crisis of human and planetary survival. Alternative interdisciplinary perspectives are needed at this time to broaden the discourse and the search for nonviolent solutions. This work is dedicated to that endeavor.

Sustainability and Well-Being seeks to contribute a synthesis of research-based social science analysis and a philosophical perspective on social and psychological transformation. Transcending traditional academic and disciplinary boundaries, it brings together research and perspectives from sociology, history, political economy, gender studies, humanistic psychology, deep ecology (which recognizes humans as part of and not separate from the Earth), and the expanding field of consciousness studies that integrates Western science and Eastern philosophy, especially Buddhism. The book also brings together perspectives from a wide range of regions and societies in the Global North and South, including traditional worldviews from disappearing indigenous and peasant communities and cutting edge, technology-based cyber networks and communities.

DOI: 10.1057/9781137308993

An integrated approach

A philosophical and political convergence has emerged in recent years around the twin concepts of sustainability and well-being, defining a sustainable world as one in which the "earth thrives and people can pursue flourishing lives."[3] Unlike conventional environmentalism, the integrated approach defines sustainability as constituting both the well-being of the human species and that of the natural world.[4] What we normally think of only as environmental problems—climate change, biodiversity loss, pollution, and so on—are also problems of human health and survival. In recent years, global environmental and social justice movements have come closer together in common struggles against the twin forces of environmental destruction and human impoverishment.

The term *sustainability* has degenerated into a cliché in common usage, but its underlying meaning and values call for serious consideration. The word is derived from the Latin *sustinere*, which means to hold up, support, or endure.[5] But how can sustainability of ecosystems and species be ensured when impermanence is the fact of life? Human and nonhuman forms of nature (including plants and animals) experience constant evolution: birth, growth, decay, death, and regeneration. Human survival requires consumption of natural resources, which in turn entails environmental change. Sustainability does not mean stasis and mere conservation but conscious development that honors balance and harmony between societal needs and the regenerative capacity of the planet's life-supporting systems. In 1987, the seminal United Nations' Brundtland Commission report, *Our Common Future*, defined sustainable development as "development that meets the needs of the present without compromising the ability of future generations to meet their own needs."[6] As this book argues, the needs of the present have to be determined in relation to both environmental sustenance and the well-being of all people (across gender, race, and class), not just elites. The book discusses the history of the class- and race-based politics of environmental movements in the West and the need to prevent a resurgence of those tendencies as economic crises and conflicts worsen today.

While social justice is a key component, human well-being is not reducible to social equity. The term *well-being* requires a broad understanding of life, one that goes beyond the quantitative, materialist

DOI: 10.1057/9781137308993

dimension to encompass the qualitative aspects of social and psychological health. There are enough natural resources and technical know-how today to meet human physiological needs if the necessary organizational changes and standards of environmental sustainability and social equity are met. With reorganization of production and distribution, there could be greater balance between the quantifiable dimensions of well-being (food, water, shelter, and health care) and what psychologists call subjective well-being, including the individual's need for love and belonging, esteem, self-actualization, and self-transcendence.[7] This book argues for a holistic approach to human development that goes beyond the reductionism that prevails in conventional economics and science, which approach complex phenomena in terms of their separate components.

Transformation of the economy

The modern global economy approaches the environment and humanity as mere resources and outlets for production and consumption. In so doing, it is destroying the natural integration of planetary life, seeking instead to reintegrate the environment and society through modern science, technology, and the market system .[8] This book argues that instead of attempting to dominate and subsume society and the environment within the logic of economic growth (Figure 1.1), the components of the economy—technology, property relations, the market, and finance— must be redesigned to serve the needs of environmental sustainability

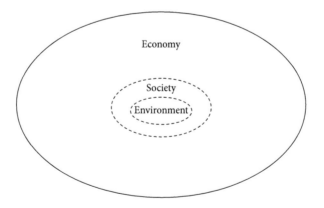

FIGURE 1.1   *Unsustainable market and technological approach*

DOI: 10.1057/9781137308993

and human well-being. The prevailing unsustainable market and technological approach and alternative sustainable approaches broadly represented by Figures 1.1, 1.2, and 1.3 in this Introduction will be elucidated in the course of the book.

The boundary between the environment and society is not distinct and tight but fluid, and the connection between them must be seen as a mutually evolving flow of energy and materials between them. Conventional approaches to sustainable development, such as the United Nations Agenda 21, promulgated by the 1992 Rio Earth Summit, approach the environment, society, and economy as three equivalent sectors, or pillars, that need to be brought into greater balance for the purposes of sustainable development (Figure 1.2).[9] But in reality these three sectors are not equivalent. The environment—planet Earth—encompasses human society and the economy within its fold (Figure 1.3). The economy, the production, and distribution of the material means of existence is only one subsystem of society.[10] The environment has primacy over the human-created spheres of society and the economy. The natural world does not need humanity for its survival, but humanity cannot survive without the environment.[11] The central idea of the ecological approach (as opposed to conventional environmentalism) is that we are part of the Earth, not apart and separate from it.[12] This does not negate the fact that in the process of adaptation and evolution humanity has made a great impact on the environment.[13]

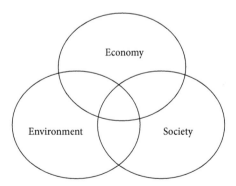

FIGURE 1.2 *Conventional approach to sustainable development*

Source: Adapted from Bob Giddings, Bill Hopwood, and Geoff O'Brien, "Environment, Economy and Society: Fitting Them Together into Sustainable Development," *Sustainable Development*, vol. 10, no. 4, 2002, pp. 187–196.

DOI: 10.1057/9781137308993

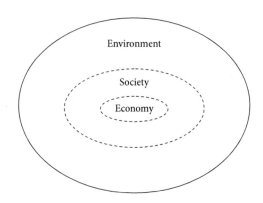

FIGURE 1.3   *Ecological approach to sustainable development*

Source: Adapted from Bob Giddings, Bill Hopwood, and Geoff O'Brien, "Environment, Economy and Society: Fitting Them Together into Sustainable Development," *Sustainable Development*, vol. 10, no. 4, 2002, pp. 187–196.

Moving beyond the cliché of "thinking outside the box," this work shows that the box is not so much the current financial system, globalization, or capitalism. Rather, it is the deeper psychological dualism of self versus other on which all other dualisms and systems of domination are constructed. Social hierarchy and domination evolved historically in conjunction with technological and material development starting in the premodern era. This book emphasizes the need to rethink binary oppositions—nature versus culture, growth versus stasis, and capitalism versus communism—to find a balanced path of human development based on interdependence and partnership between self and other instead of extreme individualism and domination. The discovery of truth cannot be assigned entirely to modern science driven by the attempt to conquer nature. The book argues that sustainability and well-being cannot be accomplished without a reorientation of the much-neglected ethical dimension in the modern economy and society. Yet, given inequalities in power and voice that characterize public forms of deliberation, it is a challenge to develop ethical reasoning that is not subservient to social conventions, religion, and unjust laws and other established institutions at both the global and local levels.[14]

Ethical transformation

Although competition, domination, and conflict seem to characterize today's world, partnership, and harmony have been present from the

DOI: 10.1057/9781137308993

beginning of human societies. These positive attributes can be the foundation of a new global ethic, as evident in the increasing integration of seemingly disparate intellectual and spiritual traditions including fields of Western science and Eastern spirituality, which uphold unity amidst diversity over reductionism and dualism. This integration can help provide the philosophical and ethical foundation to social movements for environmental sustainability and social justice. While recognizing the sources of the current crisis in unbridled economic growth and the twin forces of capital and technology, this work moves beyond popular neoliberal solutions, as well as conventional leftist critiques of corporate capitalism and ecological critiques of technology and globalization.

Western culture, like other cultures, includes many overlapping and sometimes conflicting values. Still, the hegemony of materialist and individualist thinking has prevented the West from seriously considering alternative ethical and socioeconomic pathways to a better future that can be found in non-Western cultures and Western culture itself. As the model of human progress that originated in the West threatens planetary life, as social ecologist Murray Bookchin pointed out, the world is obliged to turn to other cultures "not only for more humane values, delicate sensibilities, and richer ecological insights, but also for *technical* alternatives to our highly mystified 'powers of production.'"[15] An example is the universalist Middle Path drawn from the Buddhist teaching that informs this book.

An in-depth discussion of Buddhist philosophy is beyond the scope of this work , but basic principles of the Middle Path relevant to the global discourse on sustainability and well-being are presented here and in the final chapter. Sectarian terms such as Buddhism can, and should, be set aside and need not be a deterrent to readers from other traditions. The important concept, beyond any sectarian philosophy, is that the interdependence of self and other that underlies the balanced Middle Path can guide global social and economic development in a sustainable and equitable direction.[16]

The Middle Path

The Buddha came to the Middle Path from his own life experience, first as an heir to a royal throne living the life of sensual pleasures and later as a spiritual seeker experimenting with self-mortification. Having realized that neither extreme—overindulgence or self-denial—provided

DOI: 10.1057/9781137308993

contentment, he advocated the Middle Way of balance and moderation based on living by the "Noble Eightfold Path" that encompasses righteousness in view, intention, speech, action, livelihood, effort, mindfulness, and concentration. Identifying greed, hatred, and delusion as the roots of suffering, the Buddha promoted nonattachment to the self and cultivation of wisdom, generosity, and compassion as the foundation for human well-being.[17] While the Buddha's teaching and the Middle Path are directed toward individual liberation, they do have applicability to positive social transformation needed in the world today.

The teaching of nonviolence, tolerance, moderation, generosity, and compassion is not limited to Buddhism. Most spiritual philosophies of the world, from indigenous animist thinking and Hindu yoga to branches of Islam and Christianity, such as Sufism and Liberation Theology, uphold the basic principles and ethics of the Middle Path.[18] The ethical approach does not shy away from the realities of suffering at the individual or societal levels. It does not advocate passivity, denial, fear, fatalism, escapism, or violence in the face of contemporary environmental, social, and economic collapse. It calls for perception of reality with honesty and equanimity, exploration of the causes and evolution of current dilemmas and lessons to be learned from prevailing environmental and social movements. It encourages rational action that is respectful of self and other and present and future generations.

As scholar-monk Bhikku Bodhi writes, the Middle Path "is not a compromise between the extremes but a way that rises above them, avoiding the pitfalls into which they lead."[19] *Sustainability and Well-Being* expounds the Middle Path as an alternative to extremist ideology, whether it be monopoly capitalism, communist authoritarianism, ethno-religious fundamentalism, or a return to a romanticized premodern past. Sustainability and well-being by definition entail balance and moderation, not one-sidedness whether it be over consumption or under consumption, frenetic growth or economic stasis. The Middle Path emphasizes righteous intention: the ethics of generosity, compassion, and wisdom that can illuminate decision making on issues of production and consumption, including the adoption of appropriate technology and redistributive policies. It shows that the protection of the environment, the provision of livelihoods for people, and the alleviation of suffering of human beings and other animals would need to take priority over unregulated growth and the accumulation of wealth by a very small minority.[20] This book argues for a reorientation of values that goes

DOI: 10.1057/9781137308993

beyond cognitive learning to facilitate the shift to a balanced Middle Path. It considers the need to move away from the prevailing systems of domination toward more equitable and sustainable relations of partnership. It explores making this shift in our consciousness and in the ways we live in the here and now.

The book provides an overview of worsening environmental, social, and economic collapse (Chapter 2); the historical evolution of the domination paradigm that underlies current dilemmas (Chapter 3); lessons to be learned from environmental and social justice movements (Chapter 4); and the ethical approach and the Middle Path to sustainability and well-being (Chapter 5).[21]

Notes

1   Francis Fukuyama, *Our Posthuman Future: Consequences of the Biotechnology Revolution* (New York: Picador, 2002).

2   "The Sustainability Challenge," www.sustainabilitylabs.org (accessed Nov 12, 2012).

3   Millennium Ecosystem Assessment, *Ecosystems and Human Well-Being*, vols. 1 and 2 and Summary for Decision-Makers (Washington, DC: Island Press, 2005); "What is Sustainability?", *MIT Sloan Management Review*, http://sloanreview.mit.edu/what-is-sustainability (accessed Nov 9, 2012); Jonathon Porritt, *Capitalism as if the World Matters* (London: Earthscan, 2005), xv.

4   Porritt, *Capitalism as if the World Matters*, 319

5   The University of Redding ECIFM, "Definitions of Sustainability," http://www.ecifm.rdg.ac.uk/definitions.htm (accessed Nov 9, 2012).

6   *Our Common Future: Report of the World Commission on Environment and Development* (New York: Oxford University Press, 1987), 43.

7   Ed Diener, "Subjective Well-Being: The Science of Happiness and a Proposal for a National Index," *The American Psychologist*, 55.1 (Jan 2000): 34–43; Abraham Maslow, *The Farther Reaches of Human nature* (New York: Viking Press, 1971).

8   Murray Bookchin, *Our Synthetic Environment* (New York: Harper and Row, 1962); Al Gore, *Earth in the Balance: Ecology and the Human Spirit* (Boston: Houghton Mifflin, 1992).

9   http://web.idrc.ca/openebooks/448-2/); *Sustainable* 65th Session – General Assembly of the United Nations *www.un.org/en/ga/president/65/issues/sustdev.shtml*; Bob Giddings, Bill Hopwood, and Geoff O'Brien, "Environment, Economy and Society: Fitting Them together into Sustainable Development," *Sustainable Development*, 10.4 (2002): 187–196.

DOI: 10.1057/9781137308993

10   Karl Polanyi, *The Great Transformation* (Boston: Beacon Press, 1964).

11   Giddings, Hopwood, and O'Brien, "Environment, Economy and Society," 193.

12   Ibid., p. 193; John Seed, Joanna Macy, Pat Fleming, and Arne Naess, *Thinking Like a Mountain: Towards A Council of Beings* (Philadelphia: New Society Publishers, 1988).

13   Lynn Margulis, *Symbiosis in Cell Evolution*, 348–349, cited in Murray Bookchin, *The Ecology of Freedom: The Emergence and Dissolution of Hierarchy* (Montreal: Black Rose Books, 1991), 359.

14   Richard Paul and Linda Elder, "The Miniature Guide to Understanding the Foundations of Ethical Reasoning," Foundation for Critical Thinking, 2005, 4, www.criticalthinking.org; "Kohlberg's Stages of Moral Development," in W. C. Crain, Theories of Development (New York: Prentice-Hall, 1985), chapter 7, 118–136; Council for a Parliament of Religions, "Declaration toward a Global Ethic," www.parliamentofreligions.org/_ ... /TowardsAGlobalEthic.pdf.

15   Bookchin, *The Ecology of Freedom*, 88 (emphasis in original).

16   E. F. Schumacher, *Small is Beautiful: Economics as if People Mattered* (New York: Harper and Row, 1973).

17   Walpola Rahula, *What the Buddha Taught* (London: Gordon Fraser, 1978)

18   "Forum for Discussion of Global Issues," www.worlddialogue.org (accessed Nov 23, 2012); Sri Swami Satchidananda, *The Yoga Sutras of Patanjali* (Buckingham: Integral Yoga Publications, 1990); *i*; Georg Feurstein and Brenda Feuerstein, *Green Yoga* (Saskatchwan: Traditional Yoga Studies, 2007); Annemarie Schimmel, *Mystical Dimensions of Islam* (Chapel Hill: University of North Carolina Press, 2011); Leonardo Boff, *Ecology and Liberation: A New Paradigm* (New York: Orbis Books, 1991).

19   Bhikku Bodhi, "Tolerance and Diversity," Newsletter of the Buddhist Publication Society, no. 24, Summer-Fall 1993.

20   Chrystia Freeland, *Plutocrats: The Rise of the New Global Super-Rich and the Fall of Everyone Else* (New York: Penguin Press, 2012).

21   Following conventional usage, the book uses the terms environment and ecology interchangeably.

DOI: 10.1057/9781137308993

2
Environmental, Social, and Economic Collapse

Abstract: *This chapter provides a succinct picture of the depth, interconnectedness, and acceleration of global, environmental, and socio-economic collapse and the attendant stresses on communities. Utilizing extensive research and authoritative data, it covers environmental issues—climate change, biodiversity loss, and chemical pollution—and social and economic issues—inequality, unemployment, poverty, and militarism—and their combined impact on the environment and human well-being. It seeks to deepen awareness of global realities, their interconnections, and the urgency for change.*

Bandarage, Asoka. *Sustainability and Well-Being: The Middle Path to Environment, Society, and the Economy.* Basingstoke and New York: Palgrave Macmillan, 2013. DOI: 10.1057/9781137308993.

Human-induced environmental disasters are posing severe threats to both planetary sustainability and human well-being. This chapter provides a broad overview of the global environmental and socioeconomic collapse accompanying inequitable and unsustainable economic production and use of natural resources. It argues that the so-called "environmental problems" such as climate change, biodiversity loss, and chemical pollution are simultaneously social and economic problems with direct impact on human livelihoods, health, and survival.

Environmental collapse

Nature comprises of a shifting equilibrium. However, human activity since the Industrial Revolution has disturbed this inherent balance increasing nature's "normal" extinction rate to an extreme degree. Scientists point out that if current patterns of human activity continue, planetary biophysical systems could be destabilized, triggering "abrupt or irreversible environmental changes that would be deleterious or even catastrophic for human well-being."[1] In 2009, nearly thirty leading environmental scientists identified nine interconnected "planetary boundaries" that need to be respected if we are to avoid irreversible environmental change: climate change, stratospheric ozone, land use change, freshwater use, biological diversity, ocean acidification, nitrogen and phosphorus inputs to the biosphere and oceans, aerosol loading, and chemical pollution.[2] The study revealed that most of the variables including phosphorous inputs, ocean acidification, land use, stratospheric ozone depletion are nearing boundary values. Three boundaries—climate change, biological diversity, and nitrogen input to the biosphere—have already been crossed.

The facts of the study are startling: in 2009, there were 387 ppm (parts per million) ratio of carbon dioxide molecules to all molecules in the atmosphere (it increased to 392 ppm in 2012); a biodiversity extinction rate of more than 100 extinctions per million species; and 121 million tons of nitrogen pollution per year. These values clearly exceed the scientifically determined boundary values for sustainability: 350 ppm of atmospheric carbon dioxide, 10 extinctions per million species, and 35 million tons of nitrogen pollution annually. [3]

DOI: 10.1057/9781137308993

Climate change

The data supporting a warming climate are compelling. A 2006 study by the NASA scientists at the Goddard Institute for Space Studies led by Dr. James Hansen found that the temperature of the Earth is reaching a level "not seen in thousands of years" and that the Earth has been warming at the "remarkably rapid rate" of approximately 0.2°C per decade over the past 30 years.[4] Evidence that the warming is at least partly man made is also persuasive. The 2007 Assessment Report of the United Nations Intergovernmental Panel on Climate Change (IPPC) declared global warming to be "unequivocal," attributing it to increasing "anthropogenic greenhouse gas concentrations."[5]

Heat-trapping greenhouse gas (GHG) emissions are projected to increase 50 percent by 2050 due primarily to a 70 percent growth in energy-related CO_2 emissions. According to the *OECD Environmental Outlook*, the "global energy mix" in 2050 will not be significantly different from today: 85 percent from fossil energy (coal, oil, and natural gas); just over 10 percent from renewable energy, including biofuels; and 5 percent from nuclear energy.[6] Average global temperature is predicted to rise by 3–6°C by the end of the century, exceeding the internationally agreed goal of limiting it to 2°C above preindustrial levels.[7] Critics claim that the market-based cap and trade system, designed to reduce carbon emissions, is actually aggravating the problem by giving unfair financial advantages to major polluters and taking attention away from the search for more fundamental solutions.[8]

There is also growing scientific concern over methane (CH_4), a primary component of natural gas and an important energy source. Methane is the second most important greenhouse gas after CO_2, accounting for 16 percent of all GHG emissions resulting from human activities.[9] The Global Warming Potential (GWP) (how much heat a greenhouse gas traps in the atmosphere) of methane is expected to be more than 50 times greater than that of carbon dioxide over the next 20 years.[10] The primary source of methane is industrial animal agriculture, or factory farming, which supports global meat consumption. The production of two pounds of American steak requires "ten pounds of grain and the energy equivalent of a half a gallon of gasoline," not counting associated "soil erosion, water consumption, pesticide and fertilizer run off, ground water depletion and emissions of … methane."[11] Still, meat

DOI: 10.1057/9781137308993

production tripled during the last 40 years, rising 20 percent during the last decade and causing a host of interrelated issues including climate change, removal of land from cultivation of beans and other staple food items of poor people, pollution, and environmental, human, and animal health.[12]

A July 2010 report (based on research by 300 scientists in 48 countries) by the National Oceanic and Atmospheric Administration (NOAA) concluded that while human society evolved for thousands of years under one climatic state, a new set of "consistently warmer" climatic conditions are now taking shape. Air temperature over land and the oceans, sea surface temperature, sea level, and ocean heat are rising. Arctic sea ice, glaciers, and snow cover in the Northern hemisphere are declining.[13] The NOAA report and innumerable studies by NASA and other scientific bodies warn that owing to global warming, we are likely to see more extreme events such as severe drought, torrential rain, and violent storms in the years ahead.[14] Geologists also warn that global warming and the melting glaciers are likely to increase earthquakes, tsunamis, and volcanic eruptions.[15]

Climate change is already taking a human toll. Several South Pacific islands have already disappeared due to rising sea levels attributed to global warming. Other threatened island nations, such as the Maldives in the Indian Ocean and Tuvalu in the Pacific Ocean are making plans to relocate their populations.[16] In February 2012, the American Association for the Advancement of Science warned that 86 percent of the 226 Inuit villages in Alaska are subject to significant erosion and flooding due to warming, and many people are being forced to leave their ancestral lands.[17]

An international panel of experts writing in the journal *Science* in October 2011 called on governments and relief agencies to plan for resettlement of millions of people expected to be displaced by climate change.[18] Resettlement of "climate refugees,"—that is, victims of disasters induced by climate change—is already taking place at the rate of about 10 million people a year globally.[19] The United Nations International Organization for Migration warns that the number of climate refugees could rise to as high as 1 billion people by 2050.[20] Bangladesh, referred to as the "ground zero of climate change," is already experiencing severe flooding, destruction of agricultural land, and massive displacement of people.[21]

The carbon footprint (the relative amount of carbon a person, organization, or activity emits into the environment) of the poor, particularly

DOI: 10.1057/9781137308993

those in the Global South, is far less than that of the rich. Yet, as NASA scientists and other observers point out, the world's poor without access to resources will continue to suffer the most from climate impacts.[22] In addition, as discussed below, thousands upon thousands of other animal and plant species are also suffering the consequences of climate and related environmental changes.

Biodiversity loss

Climate change threatens the planet's biodiversity—the complex web of life, including forests and oceans, that constitutes the planetary life support system, Tropical rainforests, commonly described as the "Earth's lungs," provide the "single greatest terrestrial source of air" that sustains life.[23] While the tropical rainforests now cover just 2 percent of the Earth's land surface, they are home to two-thirds of all the living species on the planet. Human activities—principally agriculture, logging, extraction of resources, construction, and tourism—have greatly aggravated nature's normal extinction rate of rainforests. More than an acre-and-a-half of rainforest is estimated to be lost every second of every day, amounting to more than twice the size of Florida being lost every year. According to the Rainforest Action Network, if the current alarming rate of destruction continues, "half our remaining rainforests will be gone by the year 2025, and by 2060 there will be no rainforests remaining."[24]

The International Union for Conservation of Nature (IUCN) points out that current species extinction due to human impact is between 1000 and 10,000 times higher than it would be "naturally."[25] In the 2012 update of the IUCN Red List of Threatened Species, 19,817 out of 63,837 plant and animal species assessed were threatened with extinction. These included "'41 percent of amphibians, 33 percent of reef building corals, 25 percent of mammals, 13 percent of birds, and 30 percent of conifers."[26] About one-third of global freshwater biodiversity has already been extinguished, and more loss is expected in the future.[27] (The above figures may well be underestimates, given that many species have never been identified by scientists).

A Panel of scientists convened by IUCN and the International Program on the State of the Ocean reported in June 2011 that due to combined effects of global warming and other human-induced impacts, such as overfishing and nutrient run-off from farming, the world's oceans face an

DOI: 10.1057/9781137308993

"unprecedented loss of species comparable to the great mass extinctions of prehistory."[28] A United Nations report published by the Convention on Biological Diversity in May 2010 concurred that several ecosystems— the Amazon rainforest, freshwater lakes, rivers, and coral reefs—were approaching a tipping point, after which they may never recover.[29] The effects of this steady destruction on the air we breathe, the water we drink, the food we eat, and our access to new medicines and other resources are catastrophic. There is virtual unanimity among scientists that we have entered a period of mass extinction not seen since the age of dinosaurs 65 million years ago, only that this extinction is taking place at a much faster rate.[30]

Nitrogen pollution

While biodiversity loss and the "carbon footprint" are receiving increasing worldwide attention, the planetary boundary crossed by our "nitrogen footprint" and its implications for human and planetary health has received much less attention.

Since the industrial revolution, the Earth's natural nitrogen cycle has been disturbed even more than the carbon cycle. Since 1960 atmospheric carbon dioxide has increased 25 percent, whereas synthetic nitrogen has been elevated by 80 percent.[31] Human activities convert nitrogen into reactive forms that are emitted back into waterways and the atmosphere with devastating effects on climate, biodiversity, human health, and livelihoods.[32]

The single largest contributor to nitrogen pollution is the widespread use of chemical fertilizers in agriculture. Only a small part of the fertilizers is absorbed by plants. Most of the nitrogen and phosphorus runs off into rivers, lakes, and the sea, elevating toxicity thresholds and stressing aquatic ecosystems and fisheries.[33] Large amounts of animal waste discharged from concentrated animal feeding operations (CAFOs) present a serious risk to water quality. According to the US Environmental Protection Agency (EPA), states with high concentrations of CAFOs in the United States experience 20 to 30 serious water quality problems per year.[34]

Burning fossil fuels is also a major contributor to nitrogen pollution. Even the production of ethanol, a biofuel additive to gasoline, contributes to nitrogen pollution because it is derived from corn grown with massive amounts of chemical inputs. Scientific reports on biofuel production

DOI: 10.1057/9781137308993

and nitrogen emissions point out that unsustainable nitrogen outputs exacerbate ecological and social problems such as global warming, biodiversity, food and water security, and health.[35] Rachel Carson stated in *Silent Spring* over 50 years ago: "every human being is now subjected to contact with dangerous chemicals from the moment of conception until death."[36] Since then, the cycle of poison created by chemical agriculture has spread enveloping all life on earth.

Historical evidence shows that societies that fail to safeguard the environment, over consume strategic resources and exceed the sustainable carrying capacity of the environment face disease, social disequilibrium, and eventual collapse.[37]

Social collapse

In the premodern era, overconsumption among ruling classes and ecologically detrimental production methods had limited regional effects because of the small-scale and relative autonomy of economies and societies. At present, the repercussions are global. Tragically, material growth and economic and cultural integration of the world have had the most deleterious effects on the poorest communities, the very ones least responsible for climate change and environmental destruction.

Extinction

Extinction is not a phenomenon restricted to nonhuman species. Some scientists state that as a result of human dominance over the Earth we have entered a new "anthropocene" era (the current geological period, where human activities are shaping the global environment).[38] It could well be speculated that our very dominance is now moving us into a postanthropocene age of human demise.

Given the intimate connection of their livelihoods to their ecosystems, indigenous peoples worldwide are most vulnerable to climate change, destruction of biodiversity, and chemical pollution.[39] During the course of European colonization, the world's tribal population was reduced anywhere from 30 to 50 million between 1780 and 1930 due to conquest, the spread of new diseases by the Europeans, and sometimes outright extermination.[40] In 1500, there were an estimated 6 to 9 million indigenous people living in the tropical rainforests of Brazil; in 1992, less than 200,000 remained.[41] The few last remaining indigenous tribes are being

DOI: 10.1057/9781137308993

fast extinguished across the world. A cautionary example is the Boa, one of the Ten Great Tribes of the Andaman Islands east of India's mainland. In February 2010, with the death of its last remaining member, the Boa and their 65,000 year old language, which is traced to pre-Neolithic times and the first human settlement of the region by African migrants, disappeared forever.[42] Like the Boa, who faced habitat destruction, poverty, and disease, all the other Andaman tribes, especially the Jarawa, inundated by mass tourism, are also expected to disappear soon along with their languages and cultures.[43] UNESCO has reported that during the past 300 years there has been a dramatic increase in the extinction of languages and that 3000 or more languages still spoken in the world are seriously endangered.[44] A Research study from the International Diabetes Institute at Monash University in Australia also warns that if the diabetes epidemic attributed to changes in diet and lifestyle is not halted, many indigenous people—including Aborigines in Australia, the Maori in New Zealand, Pacific Islanders, and Native Americans—could become extinct by the end of the century.[45] Indigenous people are not the only ones victimized and made dispensable in the process of economic integration and cultural homogenization of the world. Many other groups are experiencing ecological and economic collapse and loss of livelihood.

Economic collapse

Livelihood

Large numbers of people have benefitted from trade, migration, and other flows that have accompanied globalization.[46] Yet, as globalization breaks down pre-existing socioeconomic systems, it is not providing alternative livelihoods and sufficient social and economic supports for large numbers of other people. The increasing volatility and collapse of financial markets in recent years have aggravated fear and uncertainty across regions and wider groups of people.

Living off the land is not a viable option for most people even those from rural agricultural communities. The sustainability of the land, livelihood, and health of farming communities are threatened by the industrialization of agriculture, especially the use of agrochemicals and biotechnology led by agribusiness corporations. The shift to new and expensive genetically modified (GM) varieties of seeds has become a

DOI: 10.1057/9781137308993

major factor in farmer indebtedness, ill health, and suicides in India.[47] Contamination of water and soil by extensive use of agrochemicals is implicated in the epidemic kidney disease and loss of livelihood of farming communities, for example, in Sri Lanka as well as Central America and Andhra Pradesh in India.[48]

During the current world economic crisis, considered to be the worst since the 1930s, unemployment and social unrest have increased in many parts of the world. The International Labour Organization warns of a generation of youth "having lost all hope of being able to work for a decent living."[49] Youth unemployment (among those between 15 and 29 years) is around 24 percent in the Middle East and North Africa;[50] Spain and Greece now have almost unthinkably high youth unemployment rates of nearly 50 percent.[51] There has been a dramatic rise in suicides associated with the financial crisis in Europe among once-successful professional and business people.[52] Unfortunately, the employment picture in the future does not bode well for Europe and much of the rest of the world. Some analysts claim that there will be more robots (machines resembling humans and capable of replicating certain human functions) than people in the workforce by the 2030s because this trend is "in the short-term interests of societal elites."[53] Some economists and social analysts have predicted that the vast majority of the 8 billion or so people expected to inhabit the earth in the first quarter of this century will be neither producers nor consumers.[54] Increasing numbers of people do not have access to means of survival including many with high educational qualifications.[55] Economy is increasingly displacing society and the environment (Figure 1.1) These problems are not attributable simply to population growth which is neither a root cause of unemployment, ecological and, social crises nor an inevitable long-term phenomenon

Current downward global trends in fertility is expected to converge to below-replacement levels by 2100 and a faster global population stabilization than earlier is expected.[56] Even with population stablization, large proportions of the world's people, both young and the old, are likely to be surplus populations mired in poverty with no apparent usefulness to the global economy. In the globalized economy, "products and services are designed, built, and marketed through global supply chains" where there is "no in and out anymore."[57] Many humans will become superfluous as capital moves toward labor displacing technology creating a world where, as social analyst Jeremy Rifkin states, there will be "factories without

DOI: 10.1057/9781137308993

workers…agricultural production without farms or farmers…Much of the global workforce…eliminated, replaced by information technology, robots, machines and biotechnology."[58] Loss of means of survival by increasing numbers of people including the well educated reflects a structural problem portending social and economic collapse.[59]

The problems of overpopulation and unemployment, like the broader problems of environmental and social destruction, are attributable largely to the technologies and the social relations of production of the global economy. Like environmental destruction, deepening global inequality is an expression of the working of a system driven by profit maximization over the broader needs of human and planetary survival. It is necessary to consider global economic inequality and its implications for sustainability and well-being.

Economic inequality

Growing economic inequality and the uneven distribution of resources and economic opportunities is a fundamental social issue of our time. According to a 2008 United Nations University study, 85 percent of all global assets belonged to the richest 10 percent of the global population, with over 50 percent of all assets being the property of the richest 2 percent. In contrast, the poorest 50 percent in the world have barely 1 percent of total global wealth.[60] Global inequality is still deepening. According to data released by Credit Suisse Research Institute in October 2010, the richest 0.5 percent of the global adult population held 35.6 percent of the world's wealth.[61] *Forbes Magazine* reported in March 2011 that the world's 1210 billionaires held a combined wealth that is more than half the total wealth of 3.01 billion adults around the world.[62]

Inequality often translates into deprivations in health, economic opportunities, and living standards for large numbers of people. [63]According to 2011 United Nations estimates, 1.44 billion people still live on less than US$1.25 a day with close to one-third of the world's population experiencing "multidimensional poverty."[64] The majority of those living in poverty are women, the primary nurturers of families and communities.[65] The international charity organization Oxfam argues that global hunger and poverty are deepened by the combined effects of climate change, depletion of natural resources, the global race to control land and water, the rush to turn food into biofuels, the growing world population, and changing diets.[66] There is also a

DOI: 10.1057/9781137308993

growing shortage of water. Water is the source of life, but two-third of the world's population is expected to suffer from lack of access to clean and safe water by 2025.[67] Food and water shortages lead to starvation, disease, and death. About 25,000 people, mostly children in the developing world die every day due to hunger or hunger-related causes.[68] Among the many diseases of poverty in the world is HIV/AIDS. In 2010, there were an estimated 1.8 million deaths from HIV/AIDS, mostly in the poverty-stricken region of Sub-Saharan Africa.[69]

Increasing concentration of global political and economic power in the hands of a small elite underlies the problem of global inequality. Trade liberalization during the last few decades has led to consolidation of transnational corporate control in practically every sector of the global economy. In the food sector, today, four companies—ADM, Bunge, Cargill, and Louis Dreyfus—account for 75 to 90 percent of the global grain trade; four other firms—Monsanto Dupont, Syngenta, and Limagrain—account for over 50 percent of global seed sales; and six corporations—DuPont, Monsanto, Syngenta, Dow, Bayer, and BASF—control 75 percent of the agrochemicals market.[70] Private suppliers dominated by the French corporations Suez and Veolia now provide water to more than 10 percent of the world's population—a figure expected to increase given the World Bank's efforts to extend the privatization of water.[71] Corporate Accountability International, a nonprofit NGO, warns that expanding private water delivery will lead to a "pay-to-play regime undermining water rights, democratic governance, and, ultimately, human security."[72] Highlighting "corporate concentration as a root cause of hunger and poverty," Oxfam warns that expected doubling of world food prices by 2030 could lead to a "permanent food crisis" affecting the poorest, and aggravating social unrest.[73]

As transnational corporations (TNCs) develop ever more sophisticated financial and technological networks, these "world empires of the 21st century"[74] control larger shares of global resources and wealth and wield more power over people's lives than most governments. A 2011 study of 43,000 transnational corporations by systems theorists at the Swiss Federal Institute of Technology in Zurich illustrates the extent of concentration: a dominant group of 147 companies with interlocking stakes control 40 percent of the monetary value in the entire network, with 737 companies controlling 80 percent of its wealth. Using network analysis, the study shows that transnational corporations "do not carry

DOI: 10.1057/9781137308993

out their business in isolation but [are] ... tied together in an extremely entangled web of control."[75] The study concludes that "a large portion of control flows to a small core of financial institutions"—among them, Barclays, JP Morgan Chase, Merrill Lynch, Credit Suisse, Goldman Sachs, and China Petrochemical Group—that can be seen as an economic "super-entity."[76]

Governments and international financial institutions such as the World Bank and IMF serve the interests of corporations by sanctioning deregulation and privatization of natural resources and public property, corporate monopolies and mergers, speculation over productive investment, state-financed infrastructure, restraints on worker salaries and benefits, and other measures.[77] Notwithstanding such institutional support, some corporate elites in both the Global North and South engage in bribery, tax evasions, and other corrupt and unethical practices harmful to society and the environment.[78]

Politicians and the corporate- and/or state-controlled media encourage competitiveness of their respective countries downplaying the threats posed by the accelerated race to control global energy and natural resources. Competitive, dualistic "us versus them" thinking contributes to nationalism, xenophobia, and ethno-religious conflict diverting attention away from common threats faced by people across cultural and national boundaries. Political fragmentation of states often serve local and external interests seeking control over territory and resources rather than the down-trodden groups that secessionist movements claim to represent.[79]

When intergovernmental and nongovernmental organizations are financially supported by corporations and powerful nation states, it is difficult for them to take independent positions pertaining to sustainability or social justice.[80] (This issue will be taken up in Chapter 4.) Critics argue that its mandate aside, the United Nations too has become an instrument to maintain the global political, economic, and military status quo. Recently 250 major civil society organizations from around the world signed a declaration calling for an end to "corporate control or cooptation of the United Nations," including the UN Convention on Climate Change.[81]

Militarization

Militarism underlies economic growth. According to the Stockholm Peace Research Institute (SIPRI), global military expenditure was US$1.62 trillion in 2010, or $236 for every person in the world, a

DOI: 10.1057/9781137308993

50 percent increase since 2001. As US energy experts point out, military activity is a "direct production component" of international trade, especially the oil trade and is "as necessary for imports as are pipelines and supertankers." [82] The regions with the greatest reserves of energy and natural resources, such as the Middle East, Africa, and Central and South Asia, have become the most militarized and conflict-ridden.[83]

The United States accounted for 41 percent of global defense expenses in 2011, followed by China (8.2 percent), Russia (4.1), UK (3.6 percent), and France (3.6 percent)(estimated figures).[84] Considered to be the "unrivalled" military and corporate power "in the history of the world," the United States maintains some 1000 military bases scattered in every region of the world.[85] According to SIPRI, between 2006 and 2011 the volume of Chinese arms exports also increased by 95 percent making China the sixth-largest supplier of arms in the world. During the same period, worldwide arms transfers increased by 24 percent. The five largest arms importers in 2007–2011 were all Asian states: India was the world's largest importer, accounting for 10 percent of global arms, followed by South Korea (6 percent), Pakistan (5 percent), China (5 percent), and the city-state of Singapore (4 percent).[86]

Military-related activities are responsible for a large proportion of natural resource consumption and environmental destruction. Heavy use of jet fuel for military activities is a major source of carbon emissions worldwide. The Pentagon is estimated to be the "largest institutional user of petroleum products and energy in general," but is exempt from all international climate agreements, including the Kyoto Protocol.[87] Energy consumption by the US Department of Defense (DOD) for 2009 was 932 trillion Btu, (British thermal unit- standard measurement stating amount of energy in fuel) which is equivalent to the energy consumed by Nigeria's population of more than 140 million.[88]

Historically, militarism has benefitted a few at the expense of the majority. The vast majority of people killed in wars today are civilians, a large proportion being women and children. According to the Information Clearing House, the number of deaths caused by the 2003 US led invasion of Iraq was 1,455,590, that is, about ten times more than estimates given in the mainstream US media.[89] Estimates based on the work of economist Joseph Stiglitz and Harvard public finance expert Linda Bilmes reveal that the war in Iraq cost the US "$720 million per

DOI: 10.1057/9781137308993

day, $500,000 per minute—enough to provide homes for nearly 6500 families, or health care for 423,529 children in just one day."[90] The wars in Iraq and Afghanistan also set precedents in the privatization and outsourcing of military intervention allowing CEOs of US defense contracting companies to acquire huge pay gains, in some cases, between 200 and 688 percent between the years 2002 and 2006.[91] This is in sharp contrast to the realities of many poor children in conflict- ridden countries recruited into armies and guerilla factions, often forcibly. According to the BBC, there are an estimated three hundred thousand child soldiers (some as young as six years old) globally, not counting growing numbers deployed as suicide bombers.[92] The current trajectory of militarism and domination is not conducive to sustainability and well-being. Historian E. P. Thompson called it the "exterminist mode of production."[93]

Technology-based defense modernization is progressing at a rapid pace. Drones, the first wave of satellite-guided unmanned aerial vehicles along with new forms of cyber and space warfare, will transform defense into a robotics arms race in space. [94]The United Nations is considering using drones for the first time to monitor the armed conflict in the Democratic Republic of Congo. [95]In addition to chemical and biological weapons, there are some 31,000 nuclear warheads deployed or in reserve in the stockpiles of just eight countries: Russia, the United States, France, the United Kingdom, China, India, Israel, and Pakistan.[96] A report from the Oxford Research Group on global threats points out that nuclear weapon modernization by existing nuclear weapons states encourages nuclear proliferation because "states such as Iran, with their perceptions of vulnerability, deem it necessary to develop their own deterrent forces."[97]

Notwithstanding the peaceful new world order envisioned at the end of the Cold War, the world is experiencing an explosion of "complex emergencies" and a multiplicity of threats. They combine terrorism, armed conflicts, and military intervention with collapse of economic, political and social institutions, environmental destruction, famine, displacement, and other human rights violations. Complex emergencies are especially evident in the Middle East, Africa, and South Asia. Communities faced with hunger, malnutrition, disease, environmental destruction, and armed conflict face varied forms of physical and social extinction: marginalization, displacement, depopulation, and disappearance.

DOI: 10.1057/9781137308993

Human suffering has existed since the beginning of our evolution as a species. The struggle to survive in the increasingly volatile and unequal global society is aggravating fear and insecurity, elevating the levels of stress and unhappiness across the world. Hyperconnectivity, instant electronic communication and speed up of the pace of life cannot meet the need for deeper human connectedness and connectedness to nature.[98] Feeling overwhelmed and powerless, many fail to recognize the social roots of suffering. They blame themselves taking out pent up frustrations and anger on each other and themselves. Fortunately, however, more and more people are waking up becoming aware of the extremism and inherent dangers of the dominant trajectory of socioeconomic development. They are recognizing the urgent need for a more balanced and compassionate relationship to the Earth, to each other, and to ourselves. Before taking up these developments, the next chapter examines the historical evolution of social domination that has brought us to the brink of disaster.

Notes

1   J. Rockström, W. Steffen, K. Noone, Å. Persson, F. S. Chapin, III, E. Lambin, T. M. Lenton, M. Scheffer, C. Folke, H. Schellnhuber, B. Nykvist, C. A. De Wit, T. Hughes, S. van der Leeuw, H. Rodhe, S. Sörlin, P. K. Snyder, R. Costanza, U. Svedin, M. Falkenmark, L. Karlberg, R. W. Corell, V. J. Fabry, J. Hansen, B. Walker, D. Liverman, K. Richardson, P. Crutzen, and J. Foley l. "Planetary Boundaries: Exploring the Safe Operating Space for Humanity." *Ecology and Society* 14.2 (2009): n.p.

2   Ibid., Abstract.

3   Will Steffen, Johan Rockström, and Robert Costanza," How Defining Planetary Boundaries Can Transform Our Approach to Growth," *Solutions*, 23, May 2011 (accessed Nov 28, 2012); see also report of United Nations Secretary-General's High-Level Panel on Global Sustainability (2012), www.un.org/gsp/report (accessed Nov 28, 2012).

4   "NASA Study Finds World Warmth Edging Ancient Levels," www.nasa.gov/vision/earth/enviornment/world_warmth.html (accessed May 16, 2012); see also, James Hansen, *Storms of My Grandchildren: The Truth About the Coming Climate Catastrophe and Our Last Chance to Save Humanity* (New York: Bloomsbury, 2009).

5   United Nations Intergovernmental Panel on Climate Change, *2007 Assessment Report of the United Nations Intergovernmental Panel on Climate Change.*

DOI: 10.1057/9781137308993

6   Organisation for Economic Co-operation and Development, *OECD Environmental Outlook,* OECD Environmental Outlook to 20502050: *The Consequences of Inaction,* www.oecd.org/environment/outlookto2050 (accessed Nov 28, 2012)

7   Ibid.

8   Tamra Gilbertson and Oscar Reyes, "Carbon Trading: How It Works and Why It Fails," *Critical Currents,* 7 (Nov 2009), The Dag Hammarskjöld Foundation, www.dhf.uu.se/publications/critical=currents/carbon-trading (accessed May 17, 2012).

9   Global Methane Initiative, www.epa.gov/globalmethane (accessed May 16, 2012).

10 United Nations Framework Convention on Climate Change, "Global Warming Potentials," http://unfccc.int/ghg-data/items/3825.php (accessed May 16, 2012); J. Hansen et al., "Air Pollutant Climate Forcings Within the Big Climate Picture," NASA Goddard Institute for Space Studies, Mar. 11, 2009, www.columbia.edu/~jeh1/2009/Copenhagen_20090311.pdf (accessed Nov 28, 2012).

11 Alan Durning, "Asking How Much Is Enough," *State of the World Report* (Washington, DC: World Watch Institute, 1991), 159.

12 Andrew Freedman, "Global Meat Production and Consumption Continue to Rise," www.worldwatch.org/global-meat-production-and-consumption-cont ... (accessed May 16, 2012); Asoka Bandarage, *Women, Population and Global Crisis: A Political-Economic Analysis* (London: Zed Books, 1997), 229.

13 National Oceanic and Atmospheric Administration, *State of the Climate in 2009,* www.noaanews.noaa.gov/stories2010/20100728_stateoftheclimate.ht.

14 Ibid; Global Warming and Climate Change Policy, www.gcmd.nasa.gov/resources/pointers/glob_warm.html Cached July 28 (accessed Nov 28, 2012).

15 Larry West, "Fire and Ice: Melting Glaciers Trigger Earthquakes, Tsunamis and Volcanoes," Environment.about.com/od.

16 Madeleine Rubenstein, "A Changing Climate for Small Island States," *State of the Planet: Blogs from the Earth Institute.* The Earth Institute at Columbia University, Dec 15, 2011; Axel Bojanowski and Christian Schwägerl, "UN Climate Body Struggling to Pinpoint Rising Sea Levels," Spiegelwww.spiegel.de/international/world/contradictory-studies-un-climate; wattsupwiththat.com/ ... /contradictory-studies-un-climate-body ... (accessed Nov 28, 2012).

17 "AAAS Coalition Explores Perspectives of Indigenous Communities on Climate Change, www.aaas.org/news/releases/2012/0206indigenous_rights.shtml (accessed Nov 28, 2012).

DOI: 10.1057/9781137308993

18   Cited in Deborah Zabarenko, "Countries Must Plan for Climate Refugees: Report," *Reuters Online*, Oct 17, 2011.

19   Ibid.

20   *Climate Refugees* "Climate Refugees Could Number *Number* 1 Billion by 2050," www.treehugger.com/ … /climate-refugees-could-number-1-billion (accessed May 17, 2012).

21   Lisa Friedman, *Climate Change* "Climate Change Makes Refugees in Bangladesh," *Scientific American, Refugees in Bangladesh* Mar 3, 2009, www. scientificamerican.com/article.cfm? … climate-change-refugees (accessed May 18, 2012); see also Michael Nash, "Climate Refugees," video documentary, PMG/Platform Group, 2010.

22   "How Will Global Warming Change Earth?", earthobservatory.nasa.gov/ Features/Global Warming/page6.php (accessed May 18, 2012).

23   Save The Rainforest Facts, *Rainforest,* www.savetherainforest.org/ savetherainforest_007.htm (accessed May 18, 2012).

24   Ibid.; *Rainforest "Cached – Similar Rates* of Rainforest Destruction and Species Loss in the Amazon," www.ecuadorexplorer.com (accessed May 18, 2012).

25   "Extinction Crisis Continues Apace," ww.iucn.org/about/work/programmes Nov 3, 2009 (accessed Nov. 29, 2012).

26   "Half of Mammals 'in Decline' Says Extinction Red List," afp.google. com/; "Extinction Crisis Continues Apace," www.iucn.org/about/work/ programmes Nov 3, 2009. IUCN 2012 update, www.wildlifeextra.com/go/ news/icun-2012.html (accessed Nov 28, 2012).

27   *OECD Environmental Outlook to 2050.*

28   Michael McCarthy, "Oceans on brink of catastrophe," *Independent Online,* June 21, 2011.

29   Matthew Knight, "I. N. Report: Eco-systems at Tipping Point," edition.cnn. com/2010/WORLD (accessed Nov 28, 2010).

30   Gary Strieker, "Scientists Agree World Faces Mass Extinction," cnn./ technology Facts—archives.cnn.com/2002/TECH/science/ … mass. extinction/index.html, Aug 23, 2002 (accessed Nov 29, 2012); Anthony D. Barnosky, et al., "Has the Earth's sixth mass extinction already arrived?", *Nature* 3, no. 471.7336 (March 2011): 51–57; for an alternative perspective, see "SciAm Video Whiffs It on This Extinction Event, Pharyngula freethoughtblogs.com/ … /sometimes-three-minutes-isnt-enough (accessed Nov 11, 2012).

31   Alan B. Townsend and Robert W. Howarth, "Fixing the Global Nitrogen Problem," *Scientific American* (Feb 2010).

32   "Nitrogen Pollution," sciencenetlinks.com/; Umair Irfan and ClimateWire, "Nitrogen Pollution Likely to Increase Under Climate Change," *Scientific American,* Feb 16, 2012. (accessed Nov 29, 2012).

DOI: 10.1057/9781137308993

33 Jeremy David Rouse, et al., "Nitrogen Pollution: An Assessment of Its Threat to Amphibian Survival," *Environment Health Perspectives,* 107.10 (Oct 1999).

34 Carrie Hribar, "Understanding Concentrated Animal Feeding Operations and Their Impact on Communities," National Association of Local Boards of Health.

35 Townsend and Howarth, "Fixing the Global Nitrogen Problem," 70.

36 Rachel Carson, *Silent Spring* (Boston: Houghton Mifflin Co., 2002), 15.

37 Jared Diamond, *Collapse: How Societies Choose to Fail or Succeed* (New York: Viking Press, 2005).

38 Axel Bojanowski and Christian Schwägerl, "The Anthropocene Debate: Do Humans Deserve Their Own Geological Era?", *Die Spiegel Online,* Jul 8, 2011.

39 "You are an Integral Part of Nature," www.unesco.org/mab/doc/iyb/faq.pdf (accessed Nov 24, 2012).

40 Bandarage, *Women, Population and Global Crisis,* 130–132.

41 Rainforest Destruction," www.csupomona.edu/~admckettrick/projects/ … / destruction.html (accessed May 21, 2012).

42 Anny Shaw, "Last Member of 65,000-Year-Old Tribe Dies, Taking One of World's Earliest Languages to the Grave." *Mail Online,* Feb 9, 2010 (accessed Feb 20, 2013).

43 "Last Member of 65,000-Year-Old Tribe Dies"; Erica Gies, "Holding On to What Was in the Andamans," *New York Times,* Sunday Travel Section, Feb 12, 2012, .12. "Two Years after Andaman Tribe Dies, Another Faces Extinction," www.survivalinternational.org/news/8049 (accessed Nov 23, 2012).

44 "Atlas on Endangered Languages," portal.unesco.org/ … /ev.php URL_ID=16548&URL_DO=DO_TOPIC (accessed May 18, 2012).

45 Jamie Pandaram, "Diabetes threatens Aborigine Extinction," www.smh.com. au/news. (accessed Nov 29, 2012).

46 Ian Goldin and Kenneth A. Reinert, *Globalization for Development: Meeting New Challenges* (New York: Oxford University Press, 2012).

47 Andrew Malone, "The GM genocide: Thousands of Indian Farmers Are Committing Suicide," www.dailymail.co.uk/news/article-1082559 (accessed Nov 29, 2012); www.monsanto.com/newsviews/Pages/india-farmer-suicides. aspx; see also, Raj Patel, "The Long Green Revolution," *The Journal of Peasant Studies,* 40.1 (2013): 1–63 and Rachel Carson, *Silent Spring* (Boston: Houghton Mifflin, 2002).

48 Asoka Bandarage, "Political Economy of Epidemic Kidney Disease in Sri Lanka" (forthcoming); Sasha Chavkin, "As Kidney Disease Kills Thousands Across Continents, Scientists Scramble for Answers," Oct 15, 2012, www. publicintegrity.org (accessed Feb 20, 2013); World Health Organization, "Investigation and Evaluation of Chronic Kidney Disease of Uncertain

DOI: 10.1057/9781137308993

Aetiology in Sri Lanka, Final Report," (unpublished doc: RD/DOC/GC/06), 2012.

49   Carla Babb, "Global Youth Unemployment at All-Time High: UN," www.thedailystar.net/newDesign/news-details.php?nid=150549. (accessed Nov 29, 2012).

50   Peter Coy, "The Youth Unemployment Bomb," www.businessweek.com/magazine/content/11_07/b (accessed Nov 29, 2012).

51   Felix Salmon, blogs.reuters.com/felix … /12/ … /the-global-youth-unemployment-crisis, Dec 22, 2011 (accessed Nov 29, 2012).

52   Kristina Chew, "Dramatic Rise in Suicides By Economic Crisis in Europe," Apr. 15, 2012, http://www.care2.com/causes/dramatic-rise-in-suicides-by-economic-crisis-in-europe.html#ixzz2 (accessed Nov 29, 2012).

53   Kevin Drum, "Chart of the Day: Our Robot Overlords Will take Over soon," *Mother Jones*, Apr. 17, 2012 (accessed Nov 28, 2012).

54   Richard Barnet, "The End of Jobs, "*Third World Resurgence*, 44 (1994): 19; Jeremy Rifkin, "New Technology and the End of Jobs," www.converge.org.nz/pirm/nutech.htm (accessed Nov 26, 2012).

55   Education Lawrence Mishel, "Education is Not the Cure for High Unemployment or for Income Inequality," (Jan 12, 2011), Economic Policy Institute, www.epi.org/ … /education_is_not_the_cure_for_high_unempl …

56   "World Population to Reach 10 Billion by 2100," http://esa.un.org/wpp/Other-Information/Press_Release_WPP2010.pdf (accessed Nov 2012); Bandarage, *Women Population and Global Crisis*, 143–149; "Global Population Of 10 Billion By 2100? – Not So Fast," yaleglobal.yale.edu/content/global-population-10-billion (accessed Nov 27, 2012); Population and Development Program, www.popdev.hampshire.edu (accessed Nov 28, 2012).

57   Thomas Friedman, "Made in the World," *New York Times*, Jan 29, 2012.

58   Quoted in Martin Khor, "World Wide Unemployment Will Reach Crisis Proportions Says Social Expert," *Third World Resurgence*, 44 (1994): 23; see also Jeremy Rifkin, *The End of Work: The Decline of the Global Labor Force and the Dawn of the Post-Market Era* (New York: G.P. Putnam's Sons, 1995).

59   Education is Not the Cure for High Unemployment or for Income Inequality).

60   "Pioneering Study Shows Richest Two Percent Own Half World Wealth," www.wider.unu.edu/events/past-events/2006-events/en_GB/ … (accessed Nov 29, 2012).

61   "The Difference More Global Equality Could Make," http://toomuchonline.org/the-difference-more-equality-could-make (accessed Nov 12, 2012); "World/GlobalInequality," http://inequality.org/global-inequality/ (accessed Nov 12, 2012).

62   Ibid. *World*.

DOI: 10.1057/9781137308993

63   Gary Fields, *Poverty, Inequality and Development*.(Cambridge: Cambridge University Press, 1980).

64   Multidimensional Poverty Index, *hdr.undp.org/en/statistics/mpi/* (accessed Nov 29, 2012).

65   Bandarage, *Women, Population and Global Crisis*, 17, 202–204; "The Feminization of Poverty," Fact Sheet no. 1, www.un.org/womenwatch/daw/followup/session/presskit/fs1.htm.

66   Felicity Lawrence, "Food Prices to Double by 2030, Oxfam Warns," *guardian.co.uk* (accessed May 20, 2012).

67   World WcP. Tomorrows, "Water Crisis: >1 Out of 6 People Lack Safe Drinking Water', drinking water," (Jul 16, 2011), www.worldculturepictorial.com/ ... /world-water-crisis-facts-shortage. (accessed Nov 29, 2012).

68   "Unger Stats," http://www.wfp.org/hunger/stats (accessed Nov 29, 2012).

69   "Global HIV and AIDS Estimates," 2009 and 2010, http://www.avert.org/worldstats.htm (accessed Nov 29, 2012).

70   Felicity Lawrence, "The Global Food Crisis: ABCD of Food – How the Multinationals Dominate Trade," Global development, Poverty matters blog, www.guardian.co.uk (accessed May 21, 2012).

71   *Water* Corporate Accountability International, "The Making of A World Water Crisis," *Corporate,* www.stopcorporateabuse.org/water-campaign (accessed May 21, 2012).

72   Ibid.

73   "The Global Food Crisis"; see also "Food Prices to Double by 2030, Oxfam Warns," *guardian.co.uk*, May 31, 2011 (accessed May 20, 2012).

74   Frederic Clairmont and John Cavanagh, "The Rise of the TNC," *Third World Resurgence,* 40 (1993): 19.

75   Stefania Vitali, James B. Glattfelder, and Stefano Battiston, "The Network of Global Corporate Control," *PLoS ONE* 6.10 (2011).

76   Ibid., Abstract and Appendix, (Supporting Information) 16, 17.

77   Rajesh Makwana, "Multinational Corporations (MNCs): Beyond The Profit Motive," in *Share the World's Resources*, 2006,www.stwr.org/multinational-corporations/multinational, *Corporate Corruption* (accessed 28 Nov. 2012).

78   *Corporate Corruption News Articles* "Corporate Corruption News Articles," *Want to Know* www.wanttoknow.info/corporatecorruptionnewsarticles (accessed Nov 29, 2012).

79   Discussed in Asoka Bandarage, *The Separatist Conflict in Sri Lanka: Terrorism, Ethnicity, Political Economy* (London: Routledge, 2009), 200–208.

80   John Clark, ed., *Globalizing Civic Engagement: Civil Society and Transnational Action* (London: Earthscan, 2003), Introduction, 1–28; see also Asoka Bandarage, "The 'Norwegian Model': Political Economy of NGO Peacemaking," *The Brown Journal of World Affairs*, 17.2 (Spring/Summer 2011): 221–242.

DOI: 10.1057/9781137308993

81 Friends of the Earth, "UN Global Compact Turns a Blind Eye to
    Corporate Malpractices," May 10, 2012 (accessed May 19, 2012); Richard,
    "United Nations Indicted for Enabling Corporate Control of Water and
    Greenwashing," www.polarisinstitute.org/united_nations_indicted_for_
    enabling_corp (accessed May 21, 2010); see also Carla Stea, "Manipulation
    of the UN Security Council in Support of the US-NATO Military Agenda,"
    Global Research (Oct 1, 2012) (accessed May 19, 2012).

82 Science Daily, "Military Greenhouse Gas Emissions: EPA Should Recognize
    Environmental Impact of Protecting Foreign Oil, Researchers Urge,"
    www.sciencedaily.com/releases/2010/07/100721121657.htm (accessed May 21,
    2010).

83 Michael T. Klare, *Resource Wars: The New Landscape of Global Conflict* (New
    York: Henry Holt, 2001).

84 "15 Countries with Highest Military Expenditures," www.sipri.
    org/ ... 15/ ... 15-countries ... highest-military-expendit ... (accessed Nov 29,
    2012).

85 James Petras and H. Veltmeyer, Quoted in Richard Wilcox, "United States
    Militarism, Global Instability and Environmental Destruction," *How many
    bases,* www9.ocn.ne.jp/*Military* (accessed May 21, 2012); Gloria Shur Bilchik,
    "Military Mystery: How Many Bases Does the US Have, Anyway?" www.
    occasionalplanet.org/ ... /military-mystery-how-many-bases-does ... (accessed
    May 21, 2012).

86 SIPRI, "Rise in International Arms Transfers Is Driven by Asian Demand,"
    Mar 19, 2012 (accessed Nov 29, 2012).

87 "Censored Media Democracy in Action," www.projectcensored.org/
    (accessed May 28, 2012); Bandarage, *Women, Population and Global crisis,*
    257–259.

88 "A Look at US Military Energy Consumption," www.oilprice.com , www.
    oilprice.com (Jun 8, 2011) (accessed Nov 28, 2012).

89 "Iraq Deaths," *Just Foreign Policy,* www.justforeignpolicy.org/ (accessed May
    21, 2012).

90 Share the World's Resources, "Ending War for Profit," www.stwr.org/global
    (accessed Nov 29, 2012).

91 Ibid.; Rod Nordland, "Risks in Afghan War Shifting to Private Contractors,"
    New York Times, Feb 12, 2012. (accessed Nov 29, 2012).

92 "Children's Rights," www.bbc.co.uk/worldservice/people/features/
    childrens ... (accessed Nov 29, 2012).

93 E. P. Thompson, "Notes on Exterminism: The Last Stage of Civilization," *New
    Left Review,* 121 (May–Jun. 1980): 3–31.

94 Alfred W. McCoy "Space Warfare and the Future of US Global Power,"
    Mother Jones, Nov 8, 2012 (accessed Feb 20, 2013)

DOI: 10.1057/9781137308993

95   "UN Wants to Use Drones in DR Congo Conflict," *Agence France Presse*, Nov 23, 2012.

96   "Nuclear Stockpiles," www.nuclearfiles.org/ … /basics/nuclear-stockpiles.htm (accessed Nov 29, 2012).

97   Chris Abbott, Paul Rogers, and John Sloboda, "Global Responses to Global Threats," Oxford Research Group, UK, Report, Jun 1. 2006.

98   See, for example, Sherry Turkle, *Alone Together: Why We Expect More from Technology and Less from Each Other* (New York: Basic Books, 2011).

DOI: 10.1057/9781137308993

3
Evolution of the Domination Paradigm

Abstract: *This chapter provides a broad historical analysis of the underlying causes of environmental and social disintegration. It explores the interconnected historical evolution of technology, material surplus, and social hierarchy, as well as the psychology of dualism and systems of domination since the transition from foraging to settled agricultural societies. It shows how the market system and technology permeate all areas of life today, including human reproduction and consciousness. The chapter poses the key question of our time: the extreme manifestation of domination and how the digital revolution and biotechnology present grave consequences for sustainability and well-being and the very meaning of human existence.*

Bandarage, Asoka. *Sustainability and Well-Being: The Middle Path to Environment, Society, and the Economy.* New York: Palgrave Macmillan, 2013. DOI: 10.1057/9781137308993.

Notwithstanding overwhelming evidence, ignorance, denial, and cynicism persist about the planetary crisis described in the previous chapter, particularly in the United States. Attributing climate change and environmental disasters simply to natural causes, many individuals still refuse to accept human responsibility. Others, sensing the depth of the crisis and feeling overwhelmed, succumb to fear and fatalism by subscribing to conspiracy theories or joining millenarian movements prophesying apocalypse. Finding it difficult to adapt to fast changing social and economic circumstances, some seek escape in hedonism and addiction. Directing their frustrations and anger at culturally different others, many become prey to the hatred and violence propagated by fundamentalist ethno-religious movements.[1]

The failures to meet global targets and agendas for change are commonly attributed to political partisanship and the lack of political will in countries of the Global North, greed and corruption of international financial institutions and transnational corporations, and complicity of the mainstream media. However, the BRIC countries (Brazil, Russia, India, and China) and the countries in the Global South pursuing the dominant model of competitive economic growth also have their shortcomings.[2] In fact, a great deal of the responsibility lies with governments rife with inefficiency and corruption and ruling elites everywhere enhancing personal interests at the expense of environmental sustainability and social well-being.[3] All of us, self-absorbed individuals pursuing narrow self-interests, are accountable for current dilemmas to varying degrees.

If we are seriously concerned about human and planetary survival, we must move beyond blaming particular groups, individuals including ourselves. Instead, we need to reflect more deeply on our uncritical internalization of today's dominant values and world view. Although current environmental and social collapse are attributable largely to the excesses of capitalist exploitation and modern technology, the roots of current crises go back to the precapitalist era and the psychological and social evolution of hierarchy and domination. Over the course of modern history, a paradigm—"a set of assumptions, concepts, values, and practices that constitutes a way of viewing reality" and guiding social action—defined by domination is ascendant at the global level.[4]

DOI: 10.1057/9781137308993

Domination

Domination, whether individual or collective, is based on a psychology of dualism—for example, mind versus matter, subject versus object, and self versus other. Dualistic thinking upholds the mistaken belief that the self is entirely separate from the other and that the well-being of the self requires subjugation and victory over, even annihilation of the other. Attachment to the self lies at the root of domination.[5] Over time, a fear-based ego consciousness gave rise to social structures of hierarchy and domination, including such dualities as humans over nature, male over female, lord over serf, whites over people of color, and so on. Of course, compassion and partnership have also been present; without them human life could not have survived.

Later sections of the book take up the need to consciously strengthen partnership as the basis for sustainability and well-being. Here, it is useful to take a brief historical overview of how human domination over nature and each other spread across the world to the point where it now appears to be moving, in Murray Bookchin's words, "beyond all human control."[6]

Historical evolution

According to most accounts, Earth, the third planet from the sun, was formed some five billion years ago, and life originated in the oceans hundreds of millions of years later. After mammals emerged on land and humanoid creatures developed the ability to use fire, weapons, and language, the hominid species—known as the Neanderthal, the ancestors of humankind—evolved some 100,000 years ago.[7] Human species emerged as the dominant and the only surviving humanoid subspecies on Earth some 40,000 years ago.[8] With the evolution of social institutions and greater control over nature, human society became gradually distinguished from animal communities.[9]

In 1974, US paleoanthropologist Donald Johanson discovered in Ethiopia the 3.2-million-year-old skeleton that is considered a transitional fossil, between an ape and human. (The fossil was named Lucy, after the Beatles song, "Lucy in the Sky with Diamonds.")[10] In 2009, an

DOI: 10.1057/9781137308993

even older fossil skeleton of a human ancestor was discovered by other scientists, also in Africa.[11] As Johanson points out, recognition of our common African ancestry has great relevance to understanding humanity and our interdependence.

> No matter what people look like or where their families have lived for generations,..., the qualities that make them human can be traced to a single location ... Each and every one of us ... regardless of the color of our skin, is an African ... Whenever we grasp a branch of the human tree, its roots go back to Africa.[12]

Evolutionary biologists and cultural anthropologists agree that humanity has spent much of its history, perhaps as much as 99 percent, as foragers.[13] Hunting and gathering were carried out as needed to satisfy the basic needs for physical survival; there was no surplus accumulation. Stasis—the absence of technological and material advancements—was the chief characteristic of hunter-gatherer life. Yet primitive people seem to have enjoyed a relaxed and well-fed existence. In contrast to the standard diet of the contemporary developed world, the diet of the !Kung of Southern Africa, for example, was "extremely low in salt, saturated fats, and carbohydrates, particularly sugar, and high in polyunsaturated oils, roughage, and vitamins and minerals."[14] The low level of technological development and control over nature necessitated cooperation and equality among band members in what has been characterized as "primitive communism."[15]

The domestication of animals and technological achievements, especially invention of the plow, led to settled agricultural societies—the Neolithic Revolution—some 10,000 years ago. With the advancement of technology (for example, irrigation infrastructure) and social surplus, larger populations could be maintained. Many societies came to be characterized by social hierarchies, causing gender, ethnic, and class inequalities. The need for an agricultural labor force necessitated pronatalist ideologies to control women's sexual and reproductive capacities.[16] Feminist scholars Gerda Lerner and Riane Eisler have argued that patriarchy—male domination over women—was the original form of social domination.[17] With the evolution of states and a class of serfs and lords, the dominant groups claimed greater rights to resources and social surplus based on heredity and religion and used institutionalized violence to maintain social control.

Although they were required to pay taxes and provide in-kind services to the state and the aristocracy, the peasants and artisans retained access

DOI: 10.1057/9781137308993

to land and control over labor processes and technology. The existence of strong community and family bonds and the acceptance of one's place in the social hierarchy also provided relative social stability and security. Land-based peasant societies maintained a congruence with and respect for the rhythms and cycles of nature and a relatively balanced and slow pace of life. Peasants, like indigenous people, treated the passage of time with a sense of submission without attempts to master or save it.[18]

Modernization

The creation of the mechanical clock in the fourteenth century marked an important psychological change in humankind's relation to nature— when what Lewis Mumford has described as a belief in an "independent world of mathematically measurable sequences ... made up of seconds and minutes" began to supersede the authority and the cycles of nature.[19] Under the influence of the scientific revolution of the seventeenth century in Europe, particularly the philosophies of Francis Bacon and Rene Descartes, the cyclical vision of life of earlier societies gave way to a linear, quantitative approach to understanding nature. In sharp contrast to the veneration of Mother Goddess by many early earth-based peoples, this new mechanistic approach came to view life and the Earth as an unruly force to be dissected into its separate parts and tamed and controlled by humans through technology.[20] This rationalist and materialist attitude laid the basis for development of Western science and technology, the bureaucratic state, and the capitalist economy.

This trajectory of development that emerged under mercantilism 500 years ago, and advanced greatly since the Industrial Revolution some 250 years later, integrated the entire world within one interconnected market and technological system. These developments undermined traditional community relations and the planet itself was reduced to a resource for exploitation. (Figure 1.1) More and more, people were pitted against each other and against the natural world.[21]

From the onset of the Industrial Revolution, human lives became increasingly restricted and regimented within the time and monetary units and identities created by the confluence of the market, modern technology, and the bureaucratic state. Historian E.P. Thompson has explained how human nature came to be restructured to suit the needs of early factory production by inculcating new valuations of

DOI: 10.1057/9781137308993

time, especially by teaching children "even in their infancy to improve every shining hour," saturating their minds "with the equation, time is money."[22] Anthropologist Pierre Bourdieu observed that in colonial Algeria many peasants viewed the haste and speedup of life that accompanied colonialism and modernization "as a lack of decorum combined with diabolical ambition." Some even referred to the clock as the "devil's mill."[23] At present, people have hardly any time to spare for others or even themselves.

Capitalist development took on a global character, linked closely with European imperialism from the beginning. Adopting ideologies of modernization and development and theories of comparative advantage, capitalism integrated self-sustaining indigenous, peasant, and regional economies into the growing global economy through appropriation of land, natural resources, and labor for export production. Monocultural agriculture, mining, and other export-based production disturbed traditional patterns of crop rotation and small-scale subsistence production that were more harmonious with the regional ecosystems and cycles of nature. While small groups of the native population prospered through their cooperation with colonialism, the majority of people became poor, indebted, and dependent on the vagaries of the global market for their sustenance.[24]

Marxist historical materialism is firmly rooted within the same paradigm of materialism and technological domination as capitalism. Marx unabashedly celebrated the power of capitalism and Western colonialism as a necessary stage leading to socialism. He wrote

> The bourgeois period of history has to create the material basis of the new world ... the development of the productive powers of man and the transformation of material production into a scientific domination of natural agencies. Bourgeois industry and commerce create [the] material conditions of a new world in the same way as geological revolutions have created the surface of the earth. When a great social revolution shall have mastered the results of the bourgeois epoch, the market of the world and the modern powers of production, and subjected them to the common control of the most advanced peoples, then only will human progress cease to resemble the hideous pagan idol, who would not drink the nectar but from the skull of the slain.[25]

There is no denying the myopia of the western conception of progress and growth based on the pursuit of "quantity and egoistic acquisition."[26] What the Europeans, including the Marxists, arrogantly saw as stagnation in

DOI: 10.1057/9781137308993

non-Western societies were frequently ecologically sound practices and rich community traits, albeit "ethically and morally incompatible with the predatory dynamism of Europeans."[27] Communism failed because it too subscribed to this dynamism, and in practice it perpetuated many of the worst aspects of capitalism, such as centralization, technological dominance, militarism, and human and environmental exploitation.

While class analysis remains invaluable, violent revolution is not the solution. Marxist-based state authoritarianism failed even as solutions to the problems of social and economic inequality. The Soviet Union was characterized not only by extreme social repression but also by one of the worst environmental records known to humankind. Environmental regulations were rarely enforced and industries were allowed to release massive amounts of pollution into the air, land, and water. As a result, the former Soviet Union continues to be "home to some of the world's most polluted places."[28] Also in Communist China, Mao's "war against nature" resulted in much environmental destruction and human suffering.[29]

Global economic integration and cultural homogenization has greatly accelerated in recent years because of the revolution in information and digital technology. With the collapse of Communism in the late 1980s, the capitalist model of development became the accepted paradigm the world over. All the dominant contemporary ideologies—whether it be liberal reformism, variants of right-wing conservatism including ethno-nationalism and religious fundamentalism, or authoritarianism—now operate within the capitalist growth paradigm, which amounts to a monolithic global "market totalitarianism" or "market fundamentalism."[30] Its ideology is sacrosanct to many, but the shift to a path of environmental sustainability and human well-being calls for rethinking its underlying values and principles (compare Figures 1.1 and 1.3).

Capitalism

In contrast to the relatively static precapitalist economies, capitalism is inherently dynamic and its promise of prosperity and freedom from the bondages of tradition and limits of nature is highly alluring. The capitalist faith is based on the assumption that competition and pursuit of self-interest will lead to prosperity and the well-being of all.[31] There is no doubt that economic growth under the capitalist model has brought forth incredible technological advances and comforts that humanity does not

DOI: 10.1057/9781137308993

want to—nor need to—give up. Yet the very creativity and dynamism of capitalism and modern technology are also exploitative and destructive of people and nature. In *The Great Transformation*, Karl Polyani warned of excessive commoditization which relegates human society to being "an accessory of the economic system." In Polanyi's view, "To allow the market mechanism to be the sole director of the fate of human beings and their natural environment…would result in the demolition of society" (Figure 1.1).[32]

Capitalism is defined by constant growth and expansion. Fuelled by private accumulation of natural and productive resources, competitive scientific and technological advancement, and market expansion, capital expands to new frontiers and appropriates all regions and sectors of life. The expansionary drive also leads to concentration of resources and other means of production and distribution as seen in the rise of transnational corporations. Technological and capital expansion constitutes a single process, often termed "technocapitalism."[33] Increasingly, it is through competitive advantage in technological know-how, even more than labor exploitation, that capital expands. Prime examples are genetic engineering and communication technology.[34]

Social, environmental, and ethical standards are not built into capitalist decision-making. Economic growth committed to profit and technological advancement dispenses people from traditional livelihoods, leaving many without alternative means of survival. Employment creation and environmental sustainability are externalities to capitalist planning and production. As British environmentalist Jonathon Porritt observes in *Capitalism as if the World Matters*, there is an "extraordinary mismatch between the power of multinationals and their contributions to global employment." In 2005, the world's 200 largest corporations accounted for 28 percent of global economic activity but employed less than 0.25 percent of the global workforce. The 500 largest companies accounted for <1 percent of the global workforce in 2005. Those statistics had not increased in over two decades.[35] "Jobless growth" and displacement of people by robots are the increasing trends in the global economy.[36]

In a perverse calculation, production that creates negative-use values and destroys ecosystems and communities (such as the arms trade and toxic dumping) are counted toward gross national product (GNP) and considered to be growth.[37] Moreover, as growth increasingly shifts to the financial sector, money chasing money electronically—"casino capitalism" in cyberspace—the economy becomes more and more a fictitious

DOI: 10.1057/9781137308993

economy divorced from the direct needs of people, as well as from industrial and agricultural production and the natural environment.[38] The 2008 banking crisis and the on-going recession resulting from lack of state regulation robbed people of their retirement funds and homes, aggravating fear and insecurity across the United States and Europe.[39] Continuation of the current global economic system and the reproduction of human and planetary life are increasingly at odds.

Market and technological control

Economic and technological expansion has brought great advances, creature comforts, medical discoveries, and human longevity, but humanity is also losing fundamental freedoms and quality of life in the process. Unbridled expansion of production requires an ideology of economic determinism that equates human well-being with material consumption. Psychologist Paul Wachtel and other analysts have pointed out that capitalist growth is built on human vulnerability and a view of human beings as inherently discontent and human desire as utterly inexhaustible.[40] Younger generations raised on corporate-controlled media and without adequate social and emotional support, know very little of life outside consumerism, individualism, and competition. As the pressure for academic grades and the competition for dwindling jobs intensify, independent critical thinking is replaced by a narrow view of intelligence and creativity valued only for what will bring success in the market.

Media promotion of consumerism is so pervasive that material consumption has become the primary means of self-expression and status. As the market values seep into all areas of life, individuals too have to "brand" themselves as products to be bought and sold in the labor market and in the market for personal relationships. There is an interesting mélange of cultures and dynamism in the globalized world. However, as all of humanity becomes absorbed into the global consumerist culture, genuine cultural pluralism representing alternative values, world views, and life styles is undermined.

There is a growing recognition that excessive consumerism does not necessarily empower people or make them any happier.[41] Even the movement for personal growth and well-being in the West incorporating yoga, meditation, and similar practices has succumbed by and large to the imperatives of individualism and consumerism.[42] And the

DOI: 10.1057/9781137308993

expanding field of "happiness studies" fails to emphasize the social and cultural basis of unhappiness in excessive individualism, competition, and the inherent loneliness in modern society.[43]

Many individuals, including those who advocate for social change, enjoy interacting with each other in virtual cyberspace. But when we do so, we also lose our emotional grounding in human relationships and our physical grounding on the earth. As traditional community, family, and human bonds weaken, individuals feel increasingly fragmented, atomized, if not numbed.[44] Practically, "every change is now global in magnitude" and at an accelerating pace. More and more people are saying, "We can't keep up anymore."[45] Indeed, as the source of information and communication shifts from the print medium to electronics, many experience a simultaneous overload of infomercials and Internet "posts and links and apps and texts."[46] Plugged into personal computers and phones, the human being becomes an extension of the machine and the market. The hunger for instant gratification, at the strike of a button, makes it difficult to maintain attention or feel satiated. Partly due to difficulties in maintaining psychological balance in the face of overstimulation, diagnosis of ADHD (attention deficit and hyperactive disorder) and dependence on pharmaceutical drugs such as Ritalin are widespread among children in the United States.[47] So, too, is the use of performance enhancing drugs among athletes.

The identity, location, and movement of every individual on the planet can be identified by GPS (Global Positioning System) and other technologies. Some 2.3 billion people are now connected to the Internet.[48] Every phone call, email, and Internet communication can be recorded and monitored.[49] These advanced modes of corporate-controlled technological control and surveillance do provide conveniences to individuals. But they can also be used in the name of national security by the state, be it an authoritarian, democratic, capitalist, or socialist or even by nonstate actors, for that matter. Following the September 11, 2011, terror attacks, civil liberties in the United States, the United Kingdom, and many other countries were weakened in the name of national security. In *Our Final Century*, British astronomer, Martin Rees warns that bio and cyber technologies could become so powerful that in the hands of a fanatic one such terrorist incident could kill a million people. He points out that such an attack would not require an organized jihadist terror network. One social misfit would suffice.[50] It brings to mind the ecoterrorist Unabomber, who carried out a murder campaign in the United States. a

DOI: 10.1057/9781137308993

few decades ago ostensibly to bring attention to threats posed to human freedom by modern technological civilization.[51]

Such fanaticism and violence do not help us to question how our growing dependence on technologies beyond our understanding and control have given rise to powerlessness, victimization, and servitude rather than the promised freedom, security, and well-being.[52] Accepting freedom and choice only at face value, most people fail to see the increasing loss of economic and cultural diversity or acknowledge our deepening insecurity and fear of each other. In *Nineteen Eighty Four* published in 1949, George Orwell talked of "reality control" and "doublethink": "To know and not to know, to be conscious of complete truthfulness while telling carefully constructed lies … to repudiate morality while laying claim to it …"[53] In *The Brave New World* published in 1931, Aldous Huxley warned how unknowingly people can come to love their oppression and adore the very technology and consumerism that undo their capacities to think.[54] That new world is almost here as can be seen in the accelerating advance of GNR technologies: Genetic engineering, Nanotechnology and Robotics.

A postnature, posthuman world?

Recent developments in bioscience and technology have led to what is aptly called a biological revolution. A symposium entitled "Alter Nature: Designing Nature—Designing Human Life—Owning Life," held in Hasselt, Belgium, in February 2011, raised issues that would make the digital revolution look like child's play compared to the biological revolution. The symposium raised grave questions: "What is a designable nature? Do we want a new nature? What kind of new nature do we want? What is a designable human? What is a new human? What kind of new human do we want? Which conceptual changes do the technological possibilities imply?"[55] Undoubtedly, these questions on manipulation and redesign of nature have implications beyond science to practically every aspect of life on earth, society, and the economy. Indeed, are we losing our human identity and survival as a species in nature? As earth-based indigenous people are extinguished from the face of the Earth, are we becoming extensions of technology physically and psychologically?

A new line of thinking called transhumanism celebrates such an eventuality. Its adherents argue that artificial intelligence and other new types

DOI: 10.1057/9781137308993

of cognitive tools can not only improve intellectual and physical capabilities but also enhance emotional well-being. Nick Bostrom, director of the Future of Humanity Institute at Oxford University and a pioneer in the field, asserts that the outcome will be a new type of "posthuman" life, with beings possessing such advanced qualities and skills that they could no longer be defined "simply as humans."[56] Bostrom writes

> We will begin to use science and technology not just to manage the world around us but to manage our own human biology as well…. The changes will be faster and more profound than the very, very slow changes that would occur over tens of thousands of years as a result of natural selection and biological evolution.[57]

Redesigning life through synthetic biology, which takes the DNA of organisms as raw materials to create wholly new life forms, has been advancing at a rapid pace. Corporate-funded scientific endeavors seek to replace natural forms of life with genetically modified (GM) animal and plant life. Labeling of GM food is already a highly contentious issue between consumers and agribusiness. Genetic modification is poised to become the norm as more and more bioengineered transgenic fruits, vegetables, trees, and animals are released into the environment, raising a host of questions about biodiversity, environmental sustainability, human health, and ethics. According to a video from Fidelity Investments promoting synthetic biology, in 50 years there could be more lab-created forms of plant and animal life on the planet than those identified in nature.[58]

There are also rapid advancements in human genetic engineering. Science is attempting to go beyond somatic gene therapy, which seeks cures for genetic diseases, to germ line engineering for genetic enhancement of human beings.[59] Many analysts of scientific developments believe that we are now close to the point where "designer babies" could be produced through cloning and manipulation of genes. Apparently, the requirements of consumer-parents who desire children with specific attributes, such as height, weight, color, athletic prowess, and even mental capacities for peace and happiness could soon be met.[60]

At the turn of the twentieth century, Social Darwinist ideologies contributed to the emergence of eugenics, the belief in the betterment of the human genetic stock through selective breeding.[61] Bioethicists and others fear that genetic engineering could lead to new forms of eugenics and heritable inequality between those who can buy genetic modification and pass on traits to their offspring and others who cannot.[62] Tom

DOI: 10.1057/9781137308993

Anthanasiou, author of *Divided Planet: The Ecology of Rich and Poor*, and Marcy Darnovsky of the Center for Genetics and Society write

> The story of an "enhanced" humanity panders to some of the least attractive tendencies of our time: techno-scientific curiosity unbounded by care for social consequence, economic culture in which we cannot draw lines of any kind, hopes for our children wrought into consumerism, and deep denial of our own mortality.[63]

In the arena of high technology, nanotechnology, involving atomic manipulation, "is to inanimate nature what biotechnology is to animate matter."[64] Nanotechnology is expected to replace current manufacturing technology with entirely new materials and devices that have much greater computation and power.[65] Although this so-called "science of the small" is marketed by governments and industry as the ultimate "techno-fix," ecologists are concerned that it would "impose a new level of energy and environmental costs."[66] The field known as cryonics, claims that nanotechnology will eventually enable tissue repair and regeneration of life of frozen human corpses.[67] The dream of human immortality is still more science fiction than science.

Some scientists and futurists, however, assume that as the skills required by technology surpass the ability of humans, the creation of "technological singularity," a greater-than-human artificial intelligence through technological means will be a necessity.[68] Nick Bostorm argues that "unlike other technologies, artificial intelligences are not merely tools. They are potentially independent agents."[69] Ray Kurzeweil, the author of *The Age of Spiritual Machines*, writes "The only way to keep pace will be for ....[human] species to merge with its technology. ... There is too little nature left to return to, and there are too many human beings. For better or worse, we're stuck with technology."[70] In an interview, in 2008, in *Scientific American*, computer pioneer David Levy predicted sex and even marriage between humans and robots by 2050.[71] Meanwhile, Kurzeweil talks not only of robots surpassing human intelligence but also possessing volition and emotion, becoming "spiritual machines": "machines derived from human thinking and surpassing humans in their capacity for experience, will claim to be conscious, and thus to be spiritual."[72]

Such developments raise profound questions for which there are no easy answers. Is the dualism between life and lifeless a human construct of the "outmoded anthropocentric structures of the mind?"[73] What does

DOI: 10.1057/9781137308993

it mean for humans to merge with machines? What does it mean to be "human?" Even if we can't answer these epistemological questions, what is apparent is that the potential for self-replication in GNR technologies could lead to unforeseen and dangerous situations beyond human control. In an article entitled "Why the Future Does not Need Us," Bill Joy, former Chief Scientist of Sun Microsystems, warned that GNR technologies are "threatening to make human beings an endangered species" and that "We must do more thinking up front if we are not to be ... surprised and shocked by the consequences of our inventions."[74] Indeed, is the world just one "great laboratory of life"[75] and is nature to be reduced to its component parts and reintegrated and dominated through technology and the market as envisaged by the likes of Thomas Bacon and Rene Descartes? Should there be nothing left to life outside the market and technology? (Figure 1.1).

The main ingredients of totalitarianism—business, technology, the state, militarism, media, and the ideologies of belligerent nationalism and ethno-religious rivalry—are merging. Yet current ideologies, whether it be conservatism, liberalism, authoritarianism, or many forms of religious fundamentalism, all seek limited, piecemeal solutions within the dominant paradigm of unbridled economic growth and the scientific conquest of nature. The materialist conception of life, both the capitalist and the communist variants, approaches nature as an inanimate object to be technologically conquered and exploited for purposes of economic growth. It overlooks the reality that a total apocalypse may not be necessary to wipe out much of humanity and the technological edifice. A few tsunamis, cyclones, and earthquakes aggravated in their ferocity by global warming may be enough.[76]

Technology and the market per se are not the problems. What is the intention—values and motivation—guiding modern market and technological advancement ? Are the deeper causes of the current crisis rooted in human greed, ego, and ultimately the fear of death and attachment to self? We shall consider these concerns and the shift to a sustainable and Middle Path in the final chapter.(Figure 5.1) Environmentalism is not a panacea. There are many shades of environmentalism. Given inseparability of environmental sustainability and human well-being, it is important to distinguish those that uphold pluralism, partnership, and democracy from those that contribute to exclusion, domination, and authoritarianism. We turn to these questions in the next chapter.

DOI: 10.1057/9781137308993

Notes

1   Asoka Bandarage, "Ethnic and Religious Tension in the World: A Political-Economic Perspective," in *Global Political Economy and the Wealth of Nations: Performance, Problems and Future Prospects*, ed. Philip O'Hara (London: Routledge, 2004).

2   Organisation for Economic Co-operation and Development, *OECD Environmental OutlookOECD Environmental Outlook to 20502050: The Consequences of Inaction*, www.oecd.org/environment/outlookto2050 (accessed Nov 29, 2012).

3   Transparency International, *Global Corruption Report: Climate Change* (London: Earthscan, 2011).

4   www.thefreedictionary.com/paradigm (accessed Nov 12, 2012).

5   Walpola Rahula, *What the Buddha Taught* (London: Gordon Fraser, 1978), 43.

6   Murray Bookchin, *Ecology of Freedom* (Montreal: Black Rose Books, 1991), 349.

7   John Seed et al., *Thinking Like a Mountain* (Philadelphia: New Society Publishers, 1988), 45–51; Ray Kurzweil, *The Age of Spiritual Machines* (Middlesex, UK: Penguin Books, 2000), 261–262.

8   Brian Thomas Swimme and Mary Evelyn Tucker, *Journey of the Universe* (New Haven: Yale University Press, 2011); Kurzweil, *The Age of Spiritual Machines*, 261–262.

9   Bookchin, *Ecology of Freedom*, xxi, xxii; Edward O. Wilson, *The Social Conquest of Earth* (New York: W. W. Norton, 2012).

10   Udaily (University of Delaware), "Anthropologist Who Discovered 'Lucy' Discusses Human Origins," human origins www.udel.edu/udaily/2011/apr/johanson-lucy-042211.htm; see also Becoming Human, www.becominghuman.org.

11   "Oldest Skeleton of Human Ancestor Found," National Geographic Daily News, news.nationalgeographic.com/news/ … /091001-oldest-human-s … (accessed Nov 29, 2012).

12   "Anthropologist who discovered 'Lucy' ".

13   Marvin Harris and Eric B. Ross, *Death, Sex and Fertility: Population Regulation in Preindustrial and Developing Societies* (New York: Columbia University Press, 1987), 21.

14   Marjorie Shostack, *Nisa: The Life and Words of a !Kung Woman* (New York: Vintage Books, 1983), 15.

15   "Communism, Primitive," http://www.Encyclopedia.com (accessed Nov 20, 2012).

16   Asoka Bandarage, *Women, Population and Global Crisis: A Political-Economic Analysis* (London: Zed Books, 1997), 119–120.

17   Gerda Lerner, *The Creation of Patriarchy* (Oxford: Oxford University Press, 1986); Riane Eisler, *Chalice and the Blade: Our History, Our Future* (San Francisco: Harper and Row, 1987).

DOI: 10.1057/9781137308993

18   Bandarage, *Women, Population and Global Crisis*, 119, 308; Helena Norberg-Hodge, *Ancient Futures: Learning From Ladakh* (San Francisco: Sierra Books, 1991).

19   Lewis Mumford, *Technics and Civilization* (New York: Harcourt, 1934), cited in Neil Postman, *Amusing Ourselves to Death: Public Discourse in the Age of Show Business* (New York: Penguin Books, 1985), 11.

20   Carolyn Merchant, *Death of Nature: Women, Ecology and the Scientific Revolution* (San Francisco: Harper and Row, 1983).

21   Murray Bookchin, *Post-Scarcity Anarchism* (Oakland, CA: AK Press, 1971), 24–25.

22   E. P.Thompson, "Time, Work-Discipline and Industrial Capitalism," *Past and Present*, 38 (Dec. 1967): 87.

23   Quoted in ibid., 59.

24   Asoka Bandarage, *Colonialism in Sri Lanka: The Political Economy of the Kandyan Highlands, 1833–1886* (Berlin: Mouton, 1983).

25   Karl Marx, "The Future Results of the British Rule in India," in *On Colonialism: Articles from the New York Tribune and Other Writings*, ed. Karl Marx and Frederick Engels (New York: International Publishers, 1972), 87.

26   Bookchin, *Ecology of Freedom*, 88.

27   Ibid., 87–88.

28   "Former Soviet republics top list of world's most polluted places, World's Most Polluted Places" www.lindsayfincher.com/former-soviet-republics-top-list-of-world's most polluted places (accessed Nov 29, 2012); Murray Feschbach and Alfred Friendly, Jr, *Ecocode in the USSR* (New York: Basic Books, 1992).

29   Judith Shapiro, *Mao's War against Nature* (New York: Cambridge University Press, 2001), preface.

30   Clive Hamilton, *Growth Fetish* (London: Pluto Press, 2003), 21; Fred Block, "Reframing the Political Battle: Market Fundamentalism vs. Moral Economy," www.longviewinstitute.org/projects/moral/sorcerersapprentice. (accessed Nov 29, 2012).

31   Adam Smith, *An Inquiry into the Nature and Causes of the Wealth of Nations* (London: Methuen, 1904); Milton Friedman, *Capitalism and Freedom* (Chicago: University of Chicago Press, 1962).

32   Karl Polyani, *The Great Transformation* (Boston: Beacon Press, 1957), 73–75.

33   Bandarage, *Women, Population and Global Crisis*, 123.

34   "MIT Prof. Robert Solow Wins Nobel Prize in Economics," *Los Angeles Times*, Oct 22, 1987, articles.latimes.com/1987 … /fi-15864_1_keynesian-economist (accessed Nov 29, 2012).

35   Jonathon Porritt, *Capitalism as If the World Matters* (London: Earthscan, 2005), 251.

DOI: 10.1057/9781137308993

36 United Nations, *Keeping the Promise: A Forward-Looking Review to Promote an Agreed Action Agenda to Achieve the Millennium Development Goals by 2015*. Report of the Secretary-General (A/64/665).

37 Naomi Klein, *Shock Doctrine* (New York: Henry Holt, 2008); Raj Patel, *The Value of Nothing* (New York: Picador, 2009).

38 Asoka Bandarage, "Victims of Development," *Women's Review of Books* 5.1 (Oct 1987); Susan Strange, *Casino Capitalism* (Manchester: Manchester University Press, 1997).

39 Jon D. Wisman and Barton Baker, "Increasing Inequality, Status Insecurity, Ideology, and the Financial Crisis of 2008," Working Papers 2009–14, American University, Department of Economics, 2009.

40 Paul L. Wachtel, *The Poverty of Affluence: A Psychological Portrait of the American Way of Life* (Philadelphia: New Society Publishers, 1989), 65.

41 Richard Layard, *Happiness: Lessons From a New Science* (New York, Penguin, 2005); Tibor Scitovsky, *The Joyless Economy: The Psychology of Human Satisfaction* (Oxford:Oxford University Press, 1992).

42 Chogyam Trungpa, *Cutting Through Spiritual Materialism* (Boston: Shambala Publications, 2002); Asoka Bandarage, "In Search of a New World Order," *Women's Studies International Forum* 14.4 (1991): 345–355; Asoka Bandarage, "All Life Is Yoga," *Woman of Power* 4 (Fall 1986): 9–83.

43 Hamilton, *Growth Fetish*, 209; Christopher Lasch, *The Culture of Narcissism* (New York: W.W. Norton, 1979), 248; Bruno Frey and Alois Stutzer, *Happiness and Economics* (Princeton: Princeton University Press, 2002).

44 Sheryl Turkle, *Alone Together* (New York: Basic Books, 2011).

45 Carolyn Myss, 'Forward' in Andrew Harvey, *The Hope: A Guide to Sacred Activism* (Carlsbad, CA: Hay House, 2009), xiv.

46 Postman, *Amusing Ourselves*, 13; Michael A. Stusser, "The Tweeting, Yelping, Flickring, Foursquaring, TripAdvising Mentality," *Shambhala Sun*, Sep 2012, 28.

47 Bronwen Hruska, "Raising the Ritalin Generation," *New York Times*, Aug 18, 2012; Institute for American Values, "Hard Wired to Connect: The New Scientific Case for Authoritative Communities," (New York: Institute for American Values, 2003).

48 Global: UN – Over Two Billion People Now Connected to Internet www.unpan.org/ ... /PublicAdministrationNews/ ... /ArticleView/ ... (accessed Nov 21, 2012).

49 "Email Monitoring New Powers to Record Every Phone Call Every Phone Call," email www.telegraph.co.uk (accessed Nov 29, 2012).

50 Martin Rees, *Our Final Century* (London: William Heinemann, 2003), 48; Porritt, *Capitalism*, 213.

51 "Interview with Ted Kaczynski, Administrative Maximum Facility Prison, Florence, Colorado, USA," *Earth First Journal*, Jun 1999. Archived from the

DOI: 10.1057/9781137308993

      original on Mar 18, 2009, http://web.archive.org/web/20090318135703/http://
      www.insurgentdesire.org.uk/tedk.htm (accessed Nov 19, 2012).

52   Lasch, *Culture of Narcissism*, 248–249; Bandarage, *Women, Population and
      Global Crisis*, 236.

53   George Orwell, *Nineteen Eighty Four* (New York: New American Library,
      1949), cited in Bandarage, *Women, Population and Global Crisis*, op.cit., 298.

54   Aldous Huxley, *Brave New World* (New York: HarperCollins, 1998); Postman,
      Amusing Ourselves, vii.

55   "Alter Nature: Designing Nature – Designing Human Life – Owning Life",
      www.z33.be/ … /alter-nature-designing-nature-designing-huma … (accessed
      Nov 29, 2012).

56   Lara Farrar, "Scientists: Humans and Machines Will Merge in Future," CNN.
      Com, http://edition.cnn.com/2008/TECH/07/14/bio.tech (accessed Nov 29,
      2012).

57   "Conference Examines Man and Machine Merging," Jul 15, 2008, www.
      dailytech.com/ … Man … /article12384.htm (accessed Nov 28, 2012).

58   "Synthetic Biology Promo Video: Shocking and Horrifying," | climate-
      connections.org/ … /synthetic-biology-promo-video-sh … (accessed Nov 29,
      2012).

59   Bill Mckibben, *Enough: Staying Human in an Engineered Age* (New York:
      Henry Holt, 2003), 9–10; N. Schichor, J. Simonet, and C. Canano, "Should
      We Allow Genetic Engineering? A Public Policy Analysis of Germline
      Enhancement," *Should We Allow Genetic Engineering* 9e.devbio.com/article.
      php?id=172 (accessed Nov 29, 2012).

60   Ibid; Mckibben, *Enough*, op.cit., 59–60: Francis Fukuyama, *Our Posthuman
      Future* (New York: Picador), 19.

61   Richard Hofstadter, *Social Darwinism in American Thought* (New York: George
      Braziller, 1959), 161–167.

62   Judith Levine, "Widening Economic and other Inequalities," in *Human
      Nature on Collision Course with Genetic Engineering*, World watch Institute,
      Nov 2012, www.worldwatch.org/human-nature-collision-course-genetic-enginee
      (accessed Nov 29, 2012).

63   Tom Anthanasiou and Marcy Darnovsky, "Turning Babies into
      Commodities," *Human Nature on Collision Course with Genetic Engineering*,
      World Watch Institute, Nov. 2012, www.worldwatch.org/human-nature-
      collision-course-genetic-enginee (accessed Nov 29, 2012).

64   Pat Mooney cited in Mckibben, *Enough*, 78–79.

65   Ibid.

66   Friends of Earth International, "Nanotechnology, Climate and Energy:
      Over-Heated Promises and Hot Air?" Nov 2010, 8, www.foe.co.uk/resource/
      reports/nanotechnology_climate.pdf (accessed Nov 29, 2012).

DOI: 10.1057/9781137308993

67 Mckibben, *Enough*, 156, 222–224, 156; James L. Halperin, *The First Immortal* (New York: Ballantine, 1998); Alcor: About *Cryonics* www.alcor.org/ AboutCryonics/index.html accessed Feb 21, 2013).

68 Vernor Vinge, "How to Survive in the Post-Human Era," www-rohan.sdsu. edu/faculty/vinge/misc/singularity.html (accessed Nov 29, 2012).

69 Nick Bostrom, "When Machines Outsmart Humans," *Futures* 35.7 (2000): 759–764.

70 Kurzweil, *Spiritual Machines*, 181–182.

71 Humans Charles Q. Choi, "Humans Marrying Robots? A Q&A with David Levy," *Scientific American*, Feb 14, 2008, *Scientific* ... www.scientificamerican. com/article.cfm? ... humans-marrying (accessed Feb 21, 2013); Charles Q. Choi, "Forecast: Sex and Marriage with Robots," Oct 12, 2007 www. livescience.com/1951-forecast-sex-marriage-robots-2050.html (accessed Nov 19, 2012);

72 Kurzweil, *Spiritual Machines*, 6, 153.

73 Seed, *Thinking Like a Mountain*, 36, 83.

74 Bill Joy, "Why the Future Does Not Need Us," *Wired*, Apr. 2000; David Lorimer, *Radical Prince: The Practical Vision of the Prince of Wales* (Edinburgh: Floris Books, 2003), 44.

75 H.M. Prince Charles of Great Britain cited in David Lorimer, *Radical Prince*, 44.

76 Georgetown University, "The Tsunami Tragedy: An Educational Forum," Tsunami Tragedy: An Educational Forum ISIM Publicationscdm15036. contentdm.oclc.org/cdm/singleitem/collection/ ... /11 (accessed Nov 29, 2012).

DOI: 10.1057/9781137308993

4

Ecological and Social Justice Movements

Abstract: *This chapter considers the historical evolution of ecological and social struggles, with an emphasis on contemporary movements. It considers differences between mainstream liberal and radical approaches, including leftist and deep ecology approaches. The chapter highlights the emergence of global environmental justice movements that address sustainability and well-being by calling for limits on corporate control of resources and technological growth. The chapter weighs the strengths and weaknesses of the different movements and the lessons to be learned in moving toward a balanced path of social and economic development.*

Bandarage, Asoka. *Sustainability and Well-Being: The Middle Path to Environment, Society, and the Economy.* Basingstoke and New York: Palgrave Macmillan, 2013. DOI: 10.1057/9781137308993.

DOI: 10.1057/9781137308993

Human beings have resisted environmental destruction, social injustice, and domination in individual and collective ways since premodern times. The Jarawa in the Andaman Islands, one of the last remaining hunter-gatherer tribes but now facing extinction, were using bows and arrows to keep intruders from the "civilized" world away from their lands until very recently.[1] In feudal agricultural societies known for violent repression, some peasants refused to till the soil or pay taxes to feudal monarchs and aristocrats, just as some women sought to control their reproductive power to suit their own needs over that of men or ruling classes.[2]

With the emergence of the modern world order, resistance to the dominant forces of capital, modern science and technology, and the managerial state took on many different forms.[3] During the early period of industrialization in England, peasants and artisans resisted the enclosure of common lands and industrial management strategies. The Luddites destroyed machines in efforts to regain control over the means of production and labor. In some ways, such efforts were simultaneous struggles for social well-being and environmental sustainability.[4] In the Americas and elsewhere, indigenous resistance to commoditization of nature by Europeans similarly constituted a dual attempt to preserve livelihood and an earth-based lifestyle and veneration of Mother Earth.[5] Likewise, some peasants continued subsistence cultivation on lands to which they had customary rights refusing to become laborers on export producing monocultural plantations opened up by the colonizers.[6] In so doing, they too upheld the ethics of both environmental sustainability and social justice.

In the modern era, environmental movements encompass a broad continuum from right-wing conservative, even fascist, groups to mainstream liberal and corporate groups to radical indigenous and ecology groups and even some ecoterrorist groups that engage in acts of violence in the name of protecting the environment. In seeking an integrated agenda for sustainability and well-being, it is important to distinguish differences between movements that uphold social justice and non-violence and those that overlook or positively work against them.

Environment versus social justice

Historically, organic worldviews have coexisted with social class, ethnic, and gender domination in many societies around the world.[7] Modern

DOI: 10.1057/9781137308993

environmentalism in the West has tended to view social and economic justice as an impediment to environmental protection. Gender and ethnic considerations were not integrated into many social change movements in the United States and Europe. In fact, the convergence of branches of socialism, anarchism, and environmentalism with class prejudice and racist, sexist, authoritarian, and even fascist tendencies has not been uncommon. Since the time of the Industrial Revolution, Thomas Malthus and his followers who identify overpopulation as the root cause of environmental destruction have called for restricting the numbers of the poor as a strategy to safeguard natural resources and the environment.[8] Support for eugenicist population control (including forced sterilization of black and poor women) and immigration control gained ground even among socialists, anarchists, and feminists in the United States during the influx of immigrants and during the Depression in the 1930s.[9]

In more recent years, US biologist Garret Hardin, known for popularizing the Malthusian "life boat ethic" and who is affiliated with Carrying Capacity Network, Population-Environment Balance, and the Federation for American Immigration Reform, has advocated reduction of food and development aid to poor countries and tighter immigration laws in the North to force the Global South to bring down population growth rates.[10] The US environmental group Earth First! is known for its Malthusian statement that famines in Africa and the AIDS epidemic are nature's mechanisms for dealing with the population explosion.[11] While the above have only been statements, the most extreme form of environmentally justified racism was implemented by Nazi eco-fascism.

Eco-fascism

The historical connection between ecology and fascism in the so-called Green Wing of the Nazi Party in Germany and the strains of "eco-fascism" in contemporary political culture, both demand serious attention. It was the German zoologist Ernst Haeckel who coined the term *ecology* and established it is as a science-based study of the interaction between organisms and the environment.[12] Haeckel was also one of the leading proponents of German racism, nationalism, and imperialism. He influenced the early Nazi movement and was later honored by the Nazi regime as a "forebear" of triumphant National Socialism.[13] Historian

DOI: 10.1057/9781137308993

Peter Staudenmaier has argued that "the Nazi movement's incorporation of environmentalist themes was a crucial factor in its rise to popularity and state power."[14]

The extensive environmental legislation implemented by the Nazi regime for reforestation, the protection of animals and plant species, and the blocking of industrial development were "ranked among the most progressive in the world at the time."[15] These policies reflected ecological attitudes and practices of those at the top of the Nazi hierarchy. Staudenmaier writes

> Hitler and Himmler were both strict vegetarians and animal lovers, attracted to nature, mysticism and homeopathic cures, and staunchly opposed to vivisection and cruelty to animals. Himmler even established experimental organic farms to grow herbs for SS medicinal purposes. And Hitler, at times, could sound like a veritable Green utopian, discussing authoritatively and in detail various renewable energy sources (including environmentally appropriate hydropower and producing natural gas from sludge) as alternatives to coal, and declaring "water, winds and tides" as the energy path of the future.[16]

The lesson of Nazi environmental policy is that environmentalism outside a social justice framework can be extremely dangerous. The "marriage of ecology and authoritarian social views" allowed eco-fascism in power to develop genocide "into a necessity" under the mantle of environmental preservation.[17] The Nazi official doctrine of restoring the "unity of blood and soil," meaning the race (*Volk*) and land (and the natural environment), was used to claim the sacred German soil exclusively for the Germanic Aryan people. The concept of blood and soil (*Blut und Boden*) was used as moral justification for colonization of Eastern Europe, mass murder of ethnic others, especially the Jews, who were castigated as a "rootless, wandering people, incapable of any true relationship with the land."[18]

As economic conditions worsen and social tensions and environmental dislocations increase, nationalist and eco-fascist ideas could again gain ground in the West. Rodolfo Bahro, former leftist radical and a celebrated leader of the German Green Party, embraced eco-fascism in the 1980s. Like the Nazis before him, he promoted a racist synthesis of ecology and politics as well as "spiritualism along New Age lines."[19] In his last years, he advocated for a "selectively repressive" "eco-dictatorship" under a "Green Adolf" to facilitate withdrawal from the apocalyptic "Megamachine" of industrial culture.[20]

DOI: 10.1057/9781137308993

Eco-fascist tendencies are not limited to the Federal Republic of Germany. Wings of the New Right in Europe and some white supremacist groups in the United States have linked racial preservation and immigration control with environmental preservation.[21] A few European environmentalists, among them the influential British ecologist Edward Goldsmith, have also been severely criticized by other theorists for sociobiological ideas advocating territorial separation of immigrant groups as "a normal and necessary feature of human cultural behaviour" for maintaining social stability.[22]

Counterpoised against eco-fascist tendencies is a growing global movement for sustainability and social justice. Led by NGOs, this democratic nonviolent movement challenges the dominant model of economic growth and the domination paradigm. (compare Figures 1.1. and 1.3)

Environmental and social justice movement

The broad-based antiglobalization protests in the Global North—such as the militant protests surrounding the World Trade Organization ministerial conference of 1999 in Seattle and the violent protests at the Group of 8 (France, Germany, Italy, Great Britain, Japan, United States, Canada, and Russia) summit in Genoa 2001—have received extensive media coverage. The World Social Forum, annual meeting of civil society organizations (led largely by NGOs from the North) held as an alternative to the annual World Economic Forum in Davos and its rallying cry "Another World is Possible" are also more widely known than the more focused struggles for social change in the Global South.[23] The Occupy Wall Street protests that began in September 2011 in New York City and spread around the country and abroad were characterized more by reaction and opposition to the status quo than a vision of an alternative. While the slogan "We are the 99 percent" brought much-needed attention to the excesses of the financial services sector and growing income inequality in the United States, the demise of the Occupy Movement was due both to state repression and lack of coherence of purpose. Journalist Chris Hedges, who took part in Occupy protests, was by and large right when he wrote that the "corporate culture has stripped us of the right to express ourselves outside of the narrow confines of the established political order."[24]

DOI: 10.1057/9781137308993

There is, however, much to learn from the persistence and commitment of thousands of ongoing civil society struggles in the Global South and among indigenous and other grassroots communities in the North. Increasingly, these local struggles are becoming connected to global networks and organizations challenging the dominant model of economic development. Indigenous people, peasants, workers, women, and other civil society actors are engaged in a wide array of efforts at the local level to maintain control over land, water, other natural resources, and livelihoods. These efforts focus on stopping deforestation and the building of massive infrastructural projects, dams, highways, and so on that threaten local ecosystems and communities in Latin America and the Amazon, India, Southeast Asia and elsewhere. In October 2012, some 30,000 people led by Ekta Parishad, a Gandhian social movement went on an eight-day march to Agra in India to protest the displacement of the poor, mostly low caste "untouchable" people and indigenous communities who often live in resource-rich areas where mining operations are causing extensive environmental damage. The march was successful in getting the government of India to agree to a ten point plan to redistribute land to the poor and the marginalized. The organizers have vowed to continue protests if progress is not made on plan implementation within six months.[25] In China too, lawsuits, petitions, and even open protests for environmental sustainability are on the rise as citizens groups engage in the "rightful resistance" available to them within the authoritarian political system.[26]

The struggle against the introduction of GMOs (genetically modified organisms) and seeds is intensifying worldwide. Farmers in the Indian state of Karnataka have organized against genetically modified (GM) seeds of transnational agribusiness companies like Monsanto. Farmers fear that dependence on expensive GM seeds and agrochemicals would displace original seed varieties and destroy the integrity of regional ecosystems and farming patterns.[27] Ecuador has introduced a constitutional prohibition against GMOs, which the biotechnology industry is seeking to undermine with the US government support.[28] The struggle over water is also accelerating across both the Global South and North. Two examples are the successful efforts against water privatization and control by the transnational company Bechtel in Cochabamba, Bolivia, and the ongoing effort in the United States to take back Native American control over water in Black Mesa, Arizona, from the Peabody Energy Company, which is engaged in coal mining.[29]

DOI: 10.1057/9781137308993

One of the most catalytic global movements today is 350.org which is focused on solving the climate crisis by reducing the amount of CO_2 in the atmosphere to 350 ppm, the threshold identified by climate scientists. The movement uses online campaigns, grassroots organizing, and mass public actions in nearly 200 countries to bring pressure to "put a price on carbon and take effective action to put a cap on its production."[30] American environmentalist Bill McKibben, the organizer of 350.org, argues that the organization of simultaneous public actions around the world and dissemination of those images via the media can create "something bigger than the sum of their parts to create real political pressure". He also points out a widely held misperception about the environmental movement:

> If you have been told...that environmentalists are rich white people, go look at these pictures. Most of them show poor brown, black, Asian, and young people, because that is the primary composition of the world's population. They are every bit as interested and concerned as the environmentalists we generally see in the United States—in many cases more so because they already are dealing directly with the effects of global warming.[31]

Although not usually identified as environmentalists, people of color, especially indigenous people, have long fought against toxic dumping, mining, deforestation, and various other environmentally destructive practices on their lands. Poor people of color have borne the brunt of the negative effects of corporate and government policies pertaining to the environment. A leaked confidential memo written by the then-chief economist at the World Bank, Lawrence Summers, in December 1991, explicitly called for export of polluting industries to the Third World on the grounds of cost effectiveness. He argued that the "demand for a clean environment for aesthetic and health reasons is likely to have very high income elasticity" and that a "given amount of health-impairing pollution should be done in the country with the lowest cost, which will be the country with the lowest wages."[32] The ecological and social destruction experienced by communities of color is a form of environmental racism. Unlike environmental activism associated with "rich white people" historically, the efforts of poor people of color, in both the Global North and South, are often simultaneous struggles for ecology and social justice—in other words, for environmental justice.[33]

Recognizing concentration of corporate control and "the intertwined root causes of social injustice, ecological destruction, and economic

DOI: 10.1057/9781137308993

domination," many environmental and social justice organizations are building bridges to "strengthen their collective efforts," such as nonviolent education campaigns, civil obedience, and legislative action at the local, national, and global levels.[34] An example of such a collective effort is a recent joint statement issued by Friends of the Earth International, La Via Campesina (International Peasant Movement), World March on Women, and other civil society organizations.[35] Their collective statement questions the call made by the heads of the UN Food and Agriculture Organization and the European Bank for Reconstruction and Development in the *Wall Street Journal* in September 2012 asking governments and NGOs to "embrace the private sector as the main engine for global food production."[36] The activist groups fear that promoting further growth of agribusiness by the United Nations and powerful international agencies will only aggravate ongoing land grabbing, destruction of peasant farming, and the environment.[37]

A further example of a recent collective initiative is the joint press release by the World Rainforest Movement, Global Justice Ecology Project, Global Forest Coalition and Biofuelwatch to the Executive Secretary of the United Nations Convention on Biological Diversity (CBD) opposing the release of GM trees into the environment.[38] Fearing the consequences of expanding industrial tree plantations, the press release criticizes the CBD for considering "only the interests of the forest industry." It opposes CBD's plans to introduce guidelines for the export of GM seeds around the world, which could displace communities, jeopardize the livelihoods of millions of people, and endanger the biodiversity of forest ecosystems. The environmental activists demand a global ban on the release of GM trees by the CBD.[39]

The influence of progressive environmental cum social movements over the last several decades is evident in how much of their concerns has been taken up in mainstream liberal policy agendas. The language of sustainability and well-being is now embraced by leading world institutions including the United Nations, the World Bank, the US Agency for International Development, and transnational corporations.

Broadening the mainstream agenda

In recent decades, many high-level international conferences and declarations have articulated the interconnections of sustainability

DOI: 10.1057/9781137308993

and well-being: the 1972 United Nations Conference on the Human Environment (the Stockholm Conference); the 1987 United Nations World Commission on Environment and Development (the Brundtland Commission); Agenda 21 and the 1992 United Nations Conference on Environment and Development (the Earth Summit).[40] (Figure 1.2) The mainstream discourse has been broadened by the work of Mahbub al Haq in founding the annual Human Development Reports (published since 1990); Amartya Sen arguing for development as freedom or capability expansion; and Joseph Stiglitz calling for democratizing globalization, as well as the work of many other economists, scholars, practitioners, and activists.[41]

The obvious inadequacies of traditional measures of wealth, such as the gross domestic product, to fully indicate social progress have led to the development of quality-of life-indices, such as the United Nations Human Development Index (HDI), which measures life expectancy, literacy, and per capita income.[42] The United Nations *Human Development Report* of 1995 also introduced the Gender-related Development Index (GDI) to measure inequality in achievement between women and men.[43] Environment, (social) equity, and economic needs the "three e's" are now frequently used as the criteria of sustainability (Figure 1.2).[44] The Happy Planet Index, developed by the New Economics Foundation in 2006, goes still further to measure sustainable well-being by incorporating global data on experienced or subjective well-being in addition to life expectancy and the ecological footprint (accounting for resource consumption).[45] At the June 2012 Rio+20 Conference, the United Nations Development Programme presented the concept of a sustainable human development index to account for the cost of human development to future generations.[46]

Successes and failures

There have been both successes and failures in translating the agendas on sustainability and well-being into significant change. The Montreal Protocol of 1987, convened by the United Nations Environment Programme (UNEP) to slow emission of CFCs (chloroflurocarbons) and reverse stratospheric ozone depletion, is a major achievement.[47] Renewable energy production is now gathering so much momentum that there is excitement over the arrival of a "renewable revolution," even

DOI: 10.1057/9781137308993

a "third industrial revolution" that could bring together new communication technology and renewable energies.[48]

The UN Milennium Development Summit in 2000 set eight Milennium Development Goals (MDGs to be reached by 2015: poverty alleviation, universal primary education, gender equality, child mortality reduction, maternal health, combat HIV/AIDS and other diseases, environmental sustainability, and partnership for development.[49] Although many of the poorest counties are still behind in reaching the MDGs, significant progress has been made globally. According to the World Bank, the number of people in the developing world living in poverty (less than $1.25 a day) fell from 52 percent in 1981 (1.94 billion people) to 22 percent (1.29 billion people) in 2008.[50] In 2012, the World Bank announced that percentage of people living below the $1.25 a day had declined in every region of the developing world.[51] According to UNICEF, maternal mortality dropped from an estimated 543,000 maternal deaths in 1990 to 287,000 in 2010, and child mortality also decreased.[52] The spread of HIV is reported to be stabilized in most regions, and more people with the disease survive longer except in Africa.[53]

Progress has been made in population stabilization notwithstanding the unprecedented size of global population, estimated to reach 8 billion in 2025. Population stabilization is attributable to a combination of factors, including reduction of poverty, improvement of women's education, and wider availability of contraceptives. Current population growth is due largely to population momentum resulting from the high proportion of people in the child bearing age rather than high fertility rates. Global fertility rate has come down significantly: it is now 2.5 children per woman, and the United Nations expects it to drop to 1.85 (i.e, below the replacement level of 2.1) by 2050.[54]

Still, the failures to mitigate environmental and social and economic collapse far outweigh the successes. Even as the climate crisis intensifies, the minimal efforts to address climate change through the Kyoto Protocol—the United Nations Framework Convention on Climate Change—is derailed by economic competitiveness among nations.[55] The United Nations Climate Change Conference in Copenhagen in 2009 failed to achieve a binding agreement from any country to limit emissions of greenhouse gases.[56] The 117 countries that endorsed the target of 350 ppm were the poorest and most vulnerable countries, not rapidly industrializing countries like India and China or the "rich, powerful, and deeply fossil-fuel addicted" countries of the Global North.[57] Developing countries want the developed countries in the North to take the lead in

DOI: 10.1057/9781137308993

sharply reducing emissions. Fearing loss of their competitive advantage, the developed countries want to move away from earlier agreed-upon targets and obligations. The United States has never been a member to the Kyoto Protocol; Canada, Japan, and Russia will not take part in the second round of the Protocol starting in 2013, and Australia and New Zealand also remain uncommitted.[58]

The Global Compact, the UN initiative to encourage corporations to adopt sustainable and socially responsible policies has also failed, by and large.[59] Even supporters of corporate-led growth lament that CSR (corporate social responsibility) has been reduced mostly to philanthropy and green marketing.[60] The United Nations Conference on Disarmament has been deadlocked for 15 years in its attempts to reduce proliferation of weapons of mass destruction.[61] The Comprehensive Nuclear Test-Ban Treaty (CTBT), first introduced in 1954 to ban atomic explosions (on the Earth's surface, atmosphere, underwater and underground), is yet to come into force. Among nuclear technology holder countries, India, North Korea, and Pakistan have yet to sign the treaty and China, Egypt, India, Iran, Israel, North Korea, Pakistan, and the United States have yet to ratify it.[62]

A key reason for the failures of mainstream agendas is the intensification of political and economic conflicts both across and within nations. These conflicts are also reflected in the polarization of the environmental movement between corporate and radical interests.

Corporate versus radical environmentalism

From the inception of the contemporary environmental movement in the United States, there has been close cooperation between large corporations and the government. According to Katherine Barkley and Steven Weissman in their 1970 article "Eco-Establishment," "elite resource planners" have put corporate access to natural resources above the interests of the environment and the consumers.[63] Upholding environmental integrity, deep ecologists and radical activists however, have long opposed the liberal market and managerial approach to environmental conservation. This is evident, for example, in the Declaration made by the Indigenous Environmental Network at the June 2012 Rio+20 Conference, which states

> We see the goals of UNCSD [United Nations Conference on Sustainable Development] Rio+20, the "Green Economy," and its premise that the world can only "save" nature by commodifying its life-giving and life-sustaining

DOI: 10.1057/9781137308993

capacities as a continuation of the colonialism that Indigenous Peoples and our Mother Earth have faced and resisted for 520 years.[64]

The UN-promoted mechanism and "pillar of the Green Economy" known as REDD (Reducing Emissions from Deforestation and Degradation), for example, was set up so that governments of the Global North could pay to stop global deforestation and claim it as part of their cuts in carbon emissions.[65] Radicals point out that the trade in carbon credits allows the world's worst polluting countries and companies to continue polluting. Groups such as The Indigenous Environment Network see REDD as a mechanism to "reap profits from evictions, land grabs, deforestation, and destruction of biodiversity."[66] Critics also point out the serious problems in carbon offsetting as a mechanism for reduction of emissions. Investigative reporter Johann Hari argues that because carbon dioxide pumped from a coal power plant remains in the atmosphere for millennia, to genuinely offset it, a forest too would have to stand for the same amount of time, an unlikely proposition.[67] But, he points out, instead of demanding an alternative to the flawed system of offsetting, the major conservation groups have been defending it.[68] The charge is also made that at the 2009 Copenhagen Climate Summit large environmental organizations failed to uphold even this flawed system by not challenging the "US and global political systems" to agree on a binding agreement on carbon emissions.[69]

A reason for this complicity seems to be the close relationship between mainstream environmental groups and corporations, a relationship built on "donations, partnerships, programs, projects, joint councils, and advisory boards" and the "lavish culture" shared by executives of the two sectors.[70] In her whistle-blowing book *Green Inc.*, journalist Christine MacDonald asserts that almost all major green organizations in the United States accept funding from companies known to cause environmental destruction. She writes

> Groups that once dedicated themselves solely to saving pandas, and parklands today compete for the favors of mining operations that remove entire mountaintops, logging and paper companies that clear-cut old—growth forests, and homebuilders who contribute to urban sprawl. They rely on funds from cruise ship companies, despite the industry's record for polluting the oceans. Among the most generous donors are the biggest environmental scofflaws of all: energy companies.[71]

DOI: 10.1057/9781137308993

McDonald shows that major conservation groups in the United States, such as The Nature Conservancy, the Conservation Fund, Conservation International, the Environmental Defense Fund, and the Natural Resoures Defense Council receive financial contributions from ExxonMobil, BP, BG Group, Chevron, Conoco Philips, and Shell Oil.[72] Extensive corporate lobbying of government and the silence of the mainstream media have enabled polluting companies to project an environmentally friendly public image. McDonald argues that contributions of the so-called "green corporations" such as Alcoa, BP, Consol, and Walmart to reducing their environmental footprints are "tiny compared to their overall negative impact on the planet."[73] It is a widely held view that oil companies are engaged in "green washing," by presenting misleading or inflated environmental credentials.[74]

"Revolving doors" exist between corporations, government, and the so-called Third Sector including philanthropic organizations at the highest levels, nationally and internationally. Dr. James Hansen the NASA scientist leading the effort to reverse climate change states that "greenwash was a near universal response of politicians to the climate change issue."[75] He points out that there were some 2340 registered energy lobbyists (not including those not registered) in Washington DC in 2009 including one former high ranking Congressman who was receiving $120,000 per quarter in 2008 (nearly half a million dollars a year) from the coal company, Peabody Energy.[76] Here is the crux of the problem, to be discussed in the next chapter: dualism of self and other.

Political and economic expediency prevails even when science is clearly showing that we are dangerously close to the climate's point of no return in global warming and that the future of humanity itself is at stake. But is wishing for a return to a romanticized premodern past that seems to underlie some strains of radical environmentalism the answer to current dilemmas?

Radical environmentalism

Radical environmental and social justice organizations such as the Grassroots Global Justice Alliance claim that they represent 99 percent of the world's people while those they call the proponents of the "false Green Economy" and the "Carbon Cowboys" espousing the market-bureaucratic approach represent only the 1 percent, the global super

DOI: 10.1057/9781137308993

elite.[77] But this reality alone is not sufficient to mobilize people to action, and it is important to recognize the limitations of radical ecological thinking and its tendency to romanticize precapitalist societies.

Undoubtedly, indigenous wisdom and new variants of ecological thinking such as deep ecology and eco-feminist movements, which view humanity as an extension of nature, are important for raising consciousness, especially when technology and the market are poised to create a supposedly postnature, posthuman world.[78] But in their uncritical and sometimes mystical approach to nature and premodernity, some radical thinkers and movements tend to overlook the deep cleavages inherent in many indigenous and peasant societies, and their relative lack of progress and personal freedom and mobility. The categorical denunciations of globalization and economic growth and total rejection of liberal market-based solutions, do not necessarily offer direction toward an alternative path of sustainable development. Given the tremendous urgency of addressing climate change and the lack of immediate alternatives, it is necessary to support state-based initiatives, such as, a carbon tax while at the same time fighting for fundamental changes in the technomaterialist world view and development model, as contradictory as it may seem.[79]

In looking to the past for inspiration, deep ecologists, many of them from the upper and middle classes, tend to overlook the yearning for change and for increased material consumption among those who are marginalized and living in subsistence communities. While a relaxed slower pace of life, contentment and a certain degree of happiness characterized many premodern societies, today more and more communities are faced with poverty, disease, environmental disasters, and war.[80] Poverty is not just a "culturally perceived" phenomenon, as some would argue.[81] As one critic of such concepts point out, the call for the "revival of subsistence economies and local traditions" as the "alternative to agribusiness" can result in "pushing most of the responsibility for solving social issues back onto the rural poor themselves."[82] While pushing women to bear the greater responsibility for change, eco-feminism too advances an essentialism that equates the female with nature and the male with culture, which constrains the potential of both sexes.[83]

While there is much to learn from the past, there is no going backward. We have to go forward toward genuine psychological and social transformation. A categorical statement, such as "only traditional societies have proved to be in any way sustainable," made by Edward Goldsmith, is not helpful.[84] Globalization is the reality. It is necessary to understand

DOI: 10.1057/9781137308993

the inherent unsustainability of the dominant model of social and economic development and the necessity for fundamental change. Critique of the existing system, however, is far easier than finding sustainable alternatives. What we need is not an attitude of antidevelopment or antiglobalization but an environmental justice movement dedicated to creating renewable and energy-efficient technologies and more just and ethical relations of production and consumption. The challenge is not the strengthening of autarkic, regional, and local solutions as much as envisioning a shift to a positive form of globalization with a multiplicity of endeavors at the individual, local, regional, and planetary levels.

The environmental and social justice movements must move beyond simply reacting to the myriad different issues of environmental and socio-economic collapse. Today we need a clear ethical vision and a way forward that avoids the extremes of both corporate unbridled growth and radical no-growth. The Middle Path offers such a vision and direction.

Notes

1   "Two Years After Andaman Tribe Dies, Another 'Faces Extinction,'" www.survivalinternational.org/news/8049, Jan 26, 2012 (accessed Nov 28, 2012).

2   Asoka Bandarage, *Women, Population and Global Crisis: A Political-Economic Analysis* (London: Zed Books, 1997), 307.

3   James C. Scott, *Weapons of the Weak: Everyday Forms of Peasant Resistance* (New Haven: Yale University Press, 1985).

4   E. P. Thompson, *The Making of the English Working Class* (City: Penguin Books, 1991); Ray Kurzeweil, *The Age of Spiritual Machines* (Middlesex, UK: Penguin, 2000), 80.

5   Lois Meyer and Benjamin Maldonado Alvarado, eds, *New World of Indigenous Resistance: Noam Chomsky and Voices from North, South, and Central America* (San Francisco: City Light Books, 2010).

6   Asoka Bandarage, *Colonialism in Sri Lanka: The Political Economy of the Kandyan Highlands, 1833–1886* (Berlin: Mouton, 1983), chap. 6.

7   Li Huey-li, "A Cross-Cultural Critique of Ecofeminism," in *Ecofeminism: Women, Animals, Nature*, ed. Greta Gaard (Philadelphia: Temple University Press, 1993).

8   Bandarage, *Women, Population and Global Crisis*, 27–30.

9   Ibid., 63–64.

10   Ibid., 35–36.

11   Ibid., 37.

DOI: 10.1057/9781137308993

12 Peter Staudenmaier, *Fascist Ecology: The "Green Wing" of the Nazi Party and its Historical Antecedents*, 16, www.spunk.org/texts/places/germany/sp001630/peter. htmin (accessed Nov 28, 2012).

13 Ibid., 4, 9.

14 Ibid., 9.

15 Raymond Dominick cited in Staudenmaier, *Fascist Ecology*, 17.

16 Ibid., 11.

17 Ibid., 4, 19, 20.

18 Ibid., 13.

19 Janet Biehel, *Ecology and the Modernization of Fascism in the German Ultra-Right*, 16, www.spunk.org/texts/places/germany/sp001630/janet.htm. (accessed Oct 22, 2012).

20 Ibid., 20–23; Regina Cochrane, "Rural Poverty and Impoverished Theory: Cultural Pluralism, Ecofeminism, and Global Justice," *Journal of Peasant Studies* 34.2 (June 2007): 191–192; Rudolf Bahro, *Avoiding Social and Ecological Disaster: The Politics of World Transformation* (Bath: Gateway Books, 1994).

21 Biehel, "Ecology and the Modernization of Fascism," 30; Cochrane, "Rural Poverty," 187–188.

22 Cochrane, "Rural Poverty," 187–188; see also Nicholas Hildyard, "'Blood' and 'Culture': Ethnic Conflict and the Authoritarian Right," Corner House Briefing 11, 29 (Jan 1999); "Black Shirts in Green Trousers," www.monbiot. com/2002/04/30/black-shirts-in-green-trousers (accessed Nov 28, 2012); Edward Goldsmith, "My answer," Jan 1, 2003. www.edwardgoldsmith.org/24/my-answer (accessed Nov 28, 2012)

23 Gunther Schonleitner, "World Social Forum: Making Another World Possible?" in *Globalizing Civic Engagement: Civil Society and Transnational Action*, ed. John Clark (London: Earthscan, 2003), 126–149; Research Unit for Political Economy "The Economic and Politics of the World Social Forum: Lessons for the Struggles Against 'Globalisation'," Aspects of India's Economy, no. 35, Research Unit For Political Economy, Mumbai, Sep. 2003 www. globalresearch.ca Jan 20, 2004 http://globalresearch.ca/articles/RUP401A. html (accessed Nov 28, 2012).

24 Chris Hedges, *Days of Destruction: Days of Revolt* (New York: Nation Books, 2012), 269.

25 "India's Peasant Farmers Gather for Protest March on Delhi," *The Guardian*, Oct 2, 2012; *marches victory Ekta Parishad* > Ekta Parishad www.ektaparishad. com (accessed Nov 28, 2012).

26 Bryan Tilt, *The Struggle for Sustainability in Rural China* (New York: Columbia University Press, 2009), preface.

27 La Via Campesina, "Convention on Biological Diversity: Farmers Demand an End to the Commercialization of Biodiveristy, GM Seeds and Synthetic

DOI: 10.1057/9781137308993

Biology," press release, Oct 11, 2012, viacampesina.org › ... › Biodiversity and Genetic Resources Cached (accessed Nov 28, 2012).

28   "Battle over *GMOs* intensifying in U.S. and abroad," Groundswell. www.groundswellinternational.org/ ... /battle-over-gmos-intensi... (accessed Nov 29, 2012).

29   Black Mesa Water Oscar Olivera, *Cochamabma Water Wars* (Boston: South End Press, 2004); Black Mesa Water Coalition, www. blackmesawatercoalition.org (accessed Nov 28, 2012).

30   Bill McKibben, "Human Flourishing Depends on What We Do Now," in *Ecologies of Human Flourishing,* ed. Donald K. Swearer and Susan Lloyd McGarry, 159 (Cambridge: Harvard University Press, 2011).

31   Ibid., 161.

32   Memorandum by Lawrence H. Summers, Chief Economist, World Bank, 12 Dec. 1991, re-printed in "Let Them Eat Pollution," *The Economist* (London), Feb 8, 1992: 66.

33   Filomina Chioma Steady, ed., *Environmental Justice in the New Millennium: Global Perspectives on Race, Ethnicity, and Human Rights,* (New York: Palgrave Macmillan, 2009) Introduction.

34   Social Justice Ecology Project, globaljusticeecology.org/ (accessed Nov 28, 2012).

35   Friends of the Earth International, www.foe.org/about-us/friends-of-the-earth-international (accessed Nov 28, 2012); La Via Campesina (International Peasant Movement) viacampesina.org/en/ (accessed Nov 28, 2012); World March of Women, www.worldmarchofwomen.org (accessed Nov 28, 2012).

36   FAO Social Watch, "FAO Accused of 'Promoting the Destruction of Peasant and Family Farming,'" Sep 20, 2012, www.socialwatch.org/node/15373 (accessed Nov 28, 2012).

37   Ibid.

38   World Rainforest Movement, www.wrm.org.uy (accessed Nov 28, 2012); *Global* Forest Coalition, globalforestcoalition.org (accessed Nov 28, 2012); Biofuelwatch, www.biofuelwatch.org.uk (accessed Nov 28, 2012).

39   Global Justice Ecology Project, Pressroom, Organizations Renew Demand to UNN for Global Ban on GM Trees, globaljusticeecology.org/pressroom. php?ID=621 (accessed Nov 28, 2012).

40   Report of the World *Commission* on Environment and Development, www. un-documents.net/wced-ocf.htm (accessed Nov 28, 2012); Agenda 21. *Agenda 21* www.un.org/esa/dsd/agenda21 (accessed Nov 28, 2012).

41   United Nations Human Development Reports, hdr.undp.org (accessed Nov 28, 2012);Amartya Sen, *Development as Freedom* (New York: Oxford University Press, 1999): Joseph Steiglitz, *Making Globalization Work* (New York: W. W. Norton, 2006).

DOI: 10.1057/9781137308993

42 Sakiko Fukuda-Parr, Kumar A. K. Shiva, *Readings in Human Development: Concepts, Measures and Policies for a Development Paradigm* (Oxford: Oxford University Press, 2005); Amartya Sen, "Human Development Index," in David Alexander Clark, ed. *The Elgar Companion To Development Studies Cheltenham* (UK: Edward Elgar Publishing, 2007); hdr.undp.org/en/statistics (accessed Sep 8, 2012); see also, A. Sen, J. E. Stiglitz and J. P. Fitoussi, *Report of the Commission on the Measurement of Economic Performance and Social Progress*, www.stiglitz-sen-fitoussi.fr/documents/rapport_anglais.pdf (accessed Sep 8, 2012).

43 Indices & Data | Composite Indices | Gender related indices (GDI ... hdr. undp.org/en/statistics/indices/gdi_gem/ (accessed Nov 28, 2012).

44 Michael Gunder, "Sustainability: Planning's Redemption or Curse?" www.planetizen.com/node/22812 *(Feb 8, 2007)* (accessed Nov 28, 2012).

45 Happy Planet Index, *Happy Planet Index* www.happyplanetindex.org (accessed Sep 11, 2012).

46 United Nations Development Programme, "Media | Press Releases | Human Development Reports," hdr.undp.org/en/mediacentre/press (accessed Nov 28, 2012).

47 "Scientific Assessment of Ozone Depletion 2010," www.esrl.noaa.gov/csd/ ... / intro.html (accessed Nov 28, 2012).

48 Sajed Kamal, *The Renewable Revolution: How We Can Fight Climate Change, Prevent Energy Wars, Revitalize the Economy and Transition to a Sustainable Future* (Earthscan, 2010); Jeremy Rifkin, *The Third Industrial Revolution* (New York: Palgrave Macmillan, 2011).

49 *Millennium Development Goals*, www.un.org/millenniumgoals (accessed Sep 8, 2012).

50 http://web.worldbank.org/WBSITE/EXTERNAL/NEWS (accessed Sep 8, 2012).

51 Poverty Analysis, Poverty Reduction and Equity, worldbank.org/WBSITE/ EXTERNAL/TOPICS?EXTPOVERTY?EXTPA/).con. (accessed Feb 21, 2013).

52 Chris Niles, "Meeting Notes Progress on Reducing Maternal and Child Mortality," Sep 25, 2012, www.unicef.org/health/index_65934.html (accessed Nov 25, 2012).

53 Combat HIV/AIDS, Malaria and Other Diseases, www.un.org/ millnniumgoals (accessed Nov 28, 2012).

54 "World Population to Reach 10 Billion by 2100," May 3, 2011, http://esa. un.org/wpp/Other-Information/Press_Release_WPP2010.pdf (accessed Nov 2012); Bandarage, Women Population and Global Crisis, 147; "Global Population Of 10 Billion By 2100? – Not So Fast," yaleglobal.yale.edu/ content/global-population-10-billion (accessed Nov 27, 2012).

DOI: 10.1057/9781137308993

55 World Development Movement, Climate Debt Report, "The End Game in Durban? How Developed Countries Bullied and Bribed to Try to Kill Kyoto," Nov. 2011; "U.S. Inaction on Climate is 'Criminal,'" Inter Press Service, Dec. 2011; "Global Warming Close to Becoming Irreversible," *Scientific American*, March 26, 2012.

56 Johann Hari, "The Wrong Kind of Green," *The Nation* Mar 4, 2010), www. thenation.com/article/wrong-kind-green (accessed Nov 28, 2012).

57 Bill McKibben, "Human Flourishing," 163.

58 Martin Khor, "Climate Talks at New Crossroads," *Climate talks at new crossroads* The Star Online, Sep 14, 2012, thestar.com.my

59 *United Nations* Complaint filed against POSCO for failure to carry out human rights due diligence," Oct 9, 2012, globalcompactcritics.blogspot.**com**/ ... / comp**laint-filed-against**- (accessed Nov 28, 2012).

60 Porritt, *Capitalism*, 242–246.

61 United Nations Office for Disarmament Affairs, Press release, 15 Sep. 2011 on 2011 Session of the Conference on Disarmament *15-Year Deadlock* www. un.org/News/Press/docs/2011/gadis3443.doc.htm (accessed Oct 10, 2012).

62 "Snail's Pace Towards Ban on Nuke Testing," IDN InDepthNews, *InDepthNews* Oct 2, 2012, www.indepthnews.info/ ... /1179-snails-pace-towards-ban-on-n ... (accessed Nov 28, 2012).

63 Katherine Barkley and Steve Weisman, "Eco-Establishment," *Ramparts*, May 1970 www.unz.org/Pub/Ramparts-1970may-00048 (accessed Nov 28, 2012).

64 Indigenous Environmental Network, "Road to Rio+20 and Beyond," http:// indigeneous4motherearthrioplus20.org/indgenous (accessed Oct 7, 2012).

65 Hari, "Wrong Kind of Green," *Nation* 3.

66 "Road to Rio=20 and Beyond," 3.

67 Hari, "Wrong Kind of Green," 7.

68 Ibid., 3.

69 Ibid, 3.

70 Christine MacDonald, *Green Inc: An Environmental Insider Reveals How a Good Cause Has Gone Bad* (Guilford: Connecticut: The Lyons Press, 2008), 28–35.

71 Ibid., xii; Dan Bacher "CEOs of Big Foundations, Environmental NGOs Rake in the 'Green,'" Aug 4, 2012, www.dailykos.com/CEOs-of-big-foundations-enviornment (accessed Nov 28, 2012); "Corporate Environmentalism," Nov 25, 2012, www.nonprofitwatch.org/edf/ (accessed Nov 28, 2012).

72 MacDonald, *Green Inc*, 99.

73 Ibid., .237.

74 Gina-Marie Cheeseman, "Oil Companies and Green Washing," May 5, 2008, *Oil Companies and Greenwashing* www.triplepundit.com/2008/ ... /oil-companies-and-greenwashi (accessed Nov 28, 2012).

DOI: 10.1057/9781137308993

75 James Hansen, *Storms of My Grandchildren: The Truth about the Coming Climate Catastrophe and Our Last Chance to Save Humanity* (New York: Bloomsbury, 2009), 224.

76 Ibid., 186.

77 "Road to Rio+20 and Beyond."

78 Seed, John, et al. *Thinking Like A Mountain.* Philadelphia: New Society, 1988; Vandana Shiva, *Staying Alive* (Boston: New Society Publishers, 1988); Vandana Shiva and Maria Mies, *Ecofeminism* (London: Zed Books, 1993).

79 Hansen, "Storms of my Grandchilkdren," op.cit., 221.

80 Helena Norberg-Hodge, *Ancient Futures, Learning From Ladakh* (San Fransisco: Sierra Books, 1991).

81 Cochrane, "Rural Poverty," 169.

82 Ibid., 172, 184, 191.

83 Ibid., 178; Li, "Cross Cultural Critique of Ecofeminism."

84 Goldsmith, "My Answer," 2.

DOI: 10.1057/9781137308993

5

Ethical Path to Sustainability and Well-Being

Abstract: *This chapter considers the ethical dimension of sustainability and well-being and the need to shift from the prevailing system of domination and extremism to a global consciousness and a social and economic system based on interdependence and partnership. It discusses the Middle Path, based on the cultivation of the ethics of moderation tolerance, nonviolence, and compassion, as the answer to the current crisis. It discusses how the values of the Middle Path can be applied at the institutional level to achieve environmental sustainability and human well-being through use of appropriate technology, rational allocation of resources, and balanced and equitable consumption.*

Bandarage, Asoka. *Sustainability and Well-Being: The Middle Path to Environment, Society, and the Economy.* Basingstoke and New York: Palgrave Macmillan, 2013. DOI: 10.1057/9781137308993.

DOI: 10.1057/9781137308993

The contemporary global crisis is more than a crisis of capitalism, a competition between capitalism and socialism, or a clash between modernity and ethno-religious fundamentalism. Our challenge today is not merely political, but human and ecological—how we conduct ourselves personally and collectively toward both the environment and each other.[1] At the root of the crisis, we face is the disjuncture between the exponential growth of the market economy and technology and the lack of an equivalent development in ethics and morality. Sustainability and well-being are fundamentally ethical issues calling for a transformation of consciousness and a stronger moral and ethical foundation.[2] Intellectual abhorrence of moralism and self-righteousness should not be allowed to deter ethical reorientation in the integration of the environment, society and the economy.

More than 200 leaders from over 40 faith traditions signed the declaration "Towards A Global Ethic" at the Parliament of the World's Religions in 1993. The declaration identified shared principles for a global ethic: nonviolence and respect for life; solidarity and a just economic order; tolerance and a life of truthfulness; and equal rights and partnership between men and women.[3] At present, there is also a growing worldwide awareness of the importance of environmental ethics. The 2000 Earth Charter, endorsed by thousands of civil-society groups and governments across six continents, expresses a consensus on the moral and ethical basis for sustainability and well-being.[4] The Charter spells out four interrelated principles: respect and care for the community of life; ecological integrity; social and economic justice; and democracy, nonviolence, and peace.[5]

According to Albert Einstein, "The significant problems we have cannot be solved at the same level of thinking with which we created them."[6] Many of the environmental and social problems we face today are attributable to the dualistic thinking underlying modern science since Descartes and other seventeenth century thinkers. The challenge now is to shift from this reductionist approach to a holistic level of thinking that calls for a broader level of awareness—consciousness—that incorporates not only the rationalist but also the affective and bodily dimensions of knowing. The importance of education that balances left (logical) and right (intuitive) hemispheres of the brain, and emotional and social intelligence is increasingly recognized today.[7] Physicist Arthur Zajonc calls for the integration of cognitive and affective learning when he writes that the "profoundly difficult task of learning to love…is also the task of learning to live in true peace and harmony with others and with nature."[8]

DOI: 10.1057/9781137308993

The global crisis has to be addressed through a shift in consciousness from mere accumulation of knowledge to a deeper integration of knowledge with compassion—wisdom. Many activists for social change, however, are cynical about calls for transformation of consciousness. Orthodox Marxists assert that "changes in consciousness and relationships presuppose changes in material conditions."[9] But materialist determinism diminishes the agency of individuals and the centrality of human transformation in bringing about social change. Ethical transformation in particular has to come from within. The opportunity for transformation is open to all—the young and old, poor and rich, white people and people of color, male and female. The depth of the current crisis requires us to move beyond a critique of capitalism to an understanding of the psychology of dualism that underlies the attitude of domination. The essential foundation of sustainability and well-being is not a mindset of separateness, competition, and domination. It is, rather, a partnership ethic.[10]

Western science tends to attribute consciousness to the brain while spiritual philosophies approach it from a deeper affective level, recognizing the limitations of rationality in grounding ethical convictions.[11] But today there is a growing convergence of branches of Western science, such as quantum physics and neuroscience, with ethical and spiritual approaches like Buddhism and yoga—an evolution that helps us understand consciousness as an integrated awareness that integrates the mind, body, and heart.[12] Modern physicists like Einstein, Niels Bohr, and David Bohm seem to speak the same language as ancient spiritual teachers like the Buddha and Patanjali (compiler of the Yoga Sutras) when they identify consciousness as energy and the "ultimate building block of everything in the universe."[13] According to Bohr, the universe is an "unbroken whole," and according to Bohm, the "whole of the universe was enfolded in each of its parts."[14] These positions question the reductionist framework of modern science of Descartes and others which sees mind and matter as separate. Not only are consciousness and matter interdependent but also new research in neuroscience indicates that consciousness can change matter. Meditation, for example, can affect brain chemistry.[15] Hence the importance of transforming consciousness as the basis for social transformation. Buddhist scholar Martin Verhoven says that the postmodern dilemma "highlights the need to reconcile facts and values, morals and machines, science with spirituality."[16] Philosopher Charlene Spretnak envisions this as "ecological postmodernism."[17] The global ethical orientation and paradigm shift we need—from domination to

DOI: 10.1057/9781137308993

partnership—can only be approached from such a deeper and holistic perspective.[18]

From domination to partnership

Domination, both individual and collective, is based on a psychology of dualism: mind versus matter, subject versus object, and self versus other. Dualisms in turn give rise to structures of social domination, such as anthropomorphism (humans over nature); patriarchy (male over female); capitalism (capital over labor); imperialism (center over periphery/North over South); feudalism (lords over serfs); white supremacy (whites over people of color); racism and ethnocentrism (races and ethnic groups over each other). Dualistic thinking cultivates the mistaken belief of the self as being entirely separate and in opposition to the other. This in turn gives rise to the erroneous notion that the well-being of the self requires domination and victory over, even annihilation of the other. At the root of domination lies egoism, excessive attachment to the self.

Today, "ego consciousness" and its ethics of individualism, domination, and competition is the driving force at the personal level as well as at the societal levels of nations, ethno-religious and gender groups, and in how humans relate toward other animal and life forms. This myopic consciousness is leading to massive destruction of the environment, widening economic disparities, and social conflicts. Hubris apart, the attempt by modern science and technology to conquer nature is driven by human existential insecurity, ultimately the fear of impermanence and death. This inner fear can be mitigated only with an alternative consciousness based on the acceptance of impermanence and mortality and the cultivation of equanimity and compassion.[19]

The alternative to ego consciousness based on a psychology of dualism is a universal consciousness based on unity within diversity. This higher consciousness understands the other as an extension of the self, and the well-being of the self and the other as inherently interdependent. It contributes to an ethic of partnership. In reality, however, both ego and universal consciousness are present within individuals and society, with one prevailing over the other depending on personal, social, and historical circumstances. Indeed, consciousness is not reducible to either ego or universalism. It is more complex and should be understood as a continuum.

DOI: 10.1057/9781137308993

Psychologist Lawrence Kohlberg has identified six stages of moral development: obedience and punishment; self-interest; conformity; authority and social order; social contract; and universal ethics.[20] These can be interpreted as levels of human consciousness ranging from the aggressive stage one—survival of the fittest and rule of physical power—to the altruistic stage six, where universal, ethical, and democratic principles prevail. While Kohlberg applied these levels and the possibilities of transformation to the individual, they can also be applied on the family, community, national, and global levels. The need is not so much for all individuals to reach universalist consciousness, as for more and more individuals and society as a whole to develop a consciousness that upholds interdependence instead of our worst fears and insecurities. What is needed is not the complete elimination of a psychology of domination but the conscious strengthening of an attitude of partnership.[21]

The good news is that a consciousness of partnership and cooperation is emerging, integrating seemingly disparate intellectual and spiritual traditions such as Buddhist, Native American, and other indigenous teachings with social justice philosophies, as well as elements of western science such as quantum theory and evolutionary biology. This synthesis is a holistic approach based on the entirety of nature rather than its separate parts. The scientific cum philosophical Gaia theory sees the Earth, Gaia, as "a self-regulating system made up from the totality of organisms" on the planet.[22] "Interbeing," a term coined by contemporary Vietnamese Buddhist scholar-monk Thich Nhat Hanh[23] also expands the definition of self to incorporate humanity as a part of the animal system, which in turn is part of the plant system and the Earth organism, and ultimately of the ever-evolving Universe. The definition of the self that psychologist Gregory Bateson described in *Steps to an Ecology of Mind*—being-in-relation-to-others—is consistent with the ontological categories of many indigenous peoples of the Americas, such as the Mayan concept "you are my other self" (*in lak eck*).[24]

The partnership paradigm embraces the reality of impermanence, the constantly changing nature of all material and mental phenomena, and the cycles of nature—birth, decay, death, and rebirth. It helps to free the human mind from attachment to narrow physical and mental entities, including the notion of a separate self, thereby developing empathetic connections with the other.[25] It is here that we find solace, equanimity, and freedom from craving and aversion. The partnership paradigm helps us to transcend the numbing of our hearts by excessive technology, materialism, and bureaucratic regimentation, and soften inside. It helps us to

DOI: 10.1057/9781137308993

feel the essential equality of all human beings and the commonality of the human and planetary community over our diversity and differences. The impulse, even the yearning, to do so lies within all of us, even the most disillusioned.. Many people experienced this as they watched the heart-breaking images of the Asian tsunami in December 2004 and the Katrina disaster in August 2005.[26] Bringing this consciousness into everyday life, not just during a crisis, is our challenge. It is in this sense that the crisis facing the world is fundamentally spiritual rather than intellectual.[27]

Transformation of the self

Accustomed to living within a competitive mechanistic world, many people today are cynical about commitment to nonviolent participatory processes. They have been taught to see life as nasty and brutish, and to believe that diversity and difference lead inevitably to conflict. Indeed, nature is not all peaceful and cooperative; conflict and violence are interwoven with harmony and interdependence. Like birth and creation, death and destruction are inevitable aspects of planetary evolution. Fighting for one's place in the hierarchy is a natural part of primate life. However, research-based studies counter the widespread belief that human violence is attributable to biologically determined aggression. They point out that aggression and war are not inevitable but are socially and historically created.[28] The pessimistic Hobbesian, Malthusian, and similar perspectives overlook the capacity of the human mind for the kind of conscious transformation that places humans above other species. Darwin himself recognized the importance of empathy and altruism in human evolution.[29] Research on compassion and altruism, which brings together scientific and spiritual investigation, is an expanding field of academic study as evident in the path-breaking work being done at a number of leading universities and academic institutes today[30] As Catholic theologian, Hans Kung has stated we need to recognize and honor our own responsibilities grounded in moral and religious commitments over "constantly appealing" to [human] rights against others.[31] The Dalai Lama also upholds individual ethical responsibility over rights when he says that

> universal humanitarianism is essential to solve global problems; compassion is the pillar of world peace;…world religions are already for world peace in this way, as are all humanitarians of whatever ideology; each individual has a universal responsibility to shape institutions to serve human needs.[32]

DOI: 10.1057/9781137308993

We need to develop an agenda for political action that moves us beyond narrow sectarianisms and economic competition between nation states and people. A universal ethical code of conduct and a nonviolent methodology as practiced by individuals like Mahatma Gandhi and Martin Luther King Jr. are needed.[33] The challenge is not to tear apart the dominant social and economic system through leftist or ethno-religious extremism but to shift to an ethical and sustainable path that upholds genuine diversity and democracy. Posing these two opposite poles as our only choice is a false construct. The Middle Path drawn from Buddhist teachings presents a guideline for moving in such a balanced trajectory. Albert Einstein stated

> The religion of the future will be cosmic religion. It should transcend a personal God and avoid dogmas and theology. Covering both the natural and the spiritual, it should be based on a religious sense arising from the experience of all things, natural and spiritual and a meaningful unity. Buddhism answers this description…If there is any religion that would cope with modern scientific needs, it would be Buddhism.[34]

The theory of cause and effect (dependent origination) in Buddhist teaching helps us understand the root causes of environmental, social and economic collapse that lie in craving, aversion and ignorance. The Middle Path in turn provides a road map for the alleviation of individual and social suffering through transformation of both the self and society. It calls for social change activism grounded on compassion, courage, and generosity instead of fear, anger, and hatred.[35]

The Middle Path

How can we apply the concept of interdependence of self and other that underlies the balanced Middle Path toward global social and economic change? Economist E.F. Schumacher envisioned an integrated approach to well-being and sustainability in what he called "Buddhist economics."[36] The terminology is not as important as the implementation of the ethical principles of sustainability and well-being: rational use of natural resources, appropriate technology, balanced consumption, more equitable distribution of wealth, and livelihoods for all.

Sustainable development

If production and consumption are not balanced with regenerative activity, the environmental, social, and economic collapse described in

DOI: 10.1057/9781137308993

Chapter 1 will worsen. If we respect planetary boundaries with regard to climate change (such as limiting carbon emissions to 350 ppm), biodiversity, and pollution, the environment can be regenerated and humans can continue to flourish on Earth.[37] Some environmentalists equate economic growth categorically with environmental degradation, taking a rigid antigrowth, conservationist position.[38] But most environmentalists today uphold limits to growth and advocate a form of sustainable Middle Path to development that honors the needs of the present while ensuring that resources are available for future generations.[39] We can learn such "transgenerational justice" from the Iroquois Confederacy (indigenous Nations of North America—Mohawks, Oneidas, Onondagas, Cayugas, Senecas, and Tuscaroras).[40] The Iroquois included what they called the Great Law—seventh generation sustainability—in their constitution to ensure that every important decision took into account its impact on the well-being of seven generations into the future.[41] The lesson from the Iroquois is that ethical, ecological, and social criteria need to be integrated into decision making at the outset rather than after environmental and social degradation has already occurred.

In a succinct definition of sustainability and ecological economics, economist Herman Daly argues that the economy should not grow beyond the ecosystem's capacity to regenerate raw material inputs into production and its capacity to absorb waste materials and energy outputs from production.[42] To do so, changes in the technologies used in production, distribution, and communication are required. Unethical production of negative use values—nuclear and conventional weapons, violent entertainment, and so on—must be replaced with production of life-enhancing goods and services. Renewable and clean sources of energy and appropriate technology based on solar, wind, and biomass must become the source of economic growth.[43] Many corporations and small businesses are successfully integrating environmental sustainability and profit making.[44] Biomimicry (from *bios* meaning life and *mimesis* meaning to imitate), a revolutionary new design discipline, studies nature and emulates nature's models and strategies to solve human problems sustainably.[45] Across the world, entrepreneurs are designing approaches to organic agriculture that adapt human needs to fit the land in contrast to agribusiness, which changes the environment through ecologically and socially damaging petrochemicals and genetic engineering.[46] (compare Figures 1.1. and 1.3) A balanced alternative is integrated pest management, which uses chemical pesticides when

DOI: 10.1057/9781137308993

necessary in conjunction with organic agricultural methods.[47] Balanced approaches are also gaining ground in other fields, especially medicine, where preventative holistic practices such as Indian ayurveda and Chinese acupuncture are being combined with western interventionist approaches.

The Middle Path does not disavow quantitative growth, the roles of corporations, or the state. Nor does it call for delinking from the global market or technology.[48] It is not possible to return to the stagnation that characterized precapitalist societies or abandon economic growth and technological innovation altogether. However, to restore both ecological and social balance, extreme homogenization and globalization need to be countered with stronger local and intermediate socioeconomic structures and cultural practices. Community-level economic systems are vital for the survival of local ecosystems, cultures, ethnic groups, communities, and livelihoods. As futurist Hazel Henderson has noted, the demand for bioregionalism and decentralization carries within it a critique of monopoly capitalism and unsustainable technological growth.[49] Bioregionalism honors local self-sufficiency as well as community control over water, land, and other natural resources, including plant and seed varieties. Calls for economic decentralization or Schumacher's "small is beautiful" approach do not entail complete autarky or abandoning the need for a global social and ecological agenda. What is required is not a complete disavowal of globalization or economic development but a more sustainable, ethical and socially responsible form of globalization that allows diversity in entrepreneurship and a more just relationship between the global and local levels of economic activity.

The Middle Path to sustainable development calls for changes in property relations, not just changes in technologies of production. Transnational corporations and nation states should be regulated in their appropriation and exploitation of natural resources, weapons production, and deployment and monetary transactions of large banks.[50] Ethical guidelines in declarations such as the Universal Declaration on Bioethics and Human Rights and the UN Declaration on Human Cloning need to be enforced vis-a-vis human germ line engineering, cloning, robotics, and nanotechnology, which, as discussed in Chapter 3, have the potential to change the very definition and meaning of human life.[51] It may be too late for a complete moratorium on patents on organisms and commercial release of genetically engineered products called for by some bioethicists on the grounds of dangers to health, biodiversity,

DOI: 10.1057/9781137308993

and ethics.[52] But it is not too late to create and enforce guidelines to distinguish between somatic genetic therapy and germ line manipulation. Both serve corporate profit. But the former seeks to cure diseases while the latter is driven by enhancement of human physical and psychological attributes.[53] The Middle Path and other religious teachings encourage us to question the ethical intentions of technologies. Are they motivated to promote wisdom, compassion, and generosity or hubris, greed, and hatred? Do they generate tolerance and moderation or domination and new forms of inequality?

Balanced consumption

As Schumacher explained, the Buddhist Middle Way is by no means "antagonistic to physical well-being. It is not wealth that stands in the way of liberation but the attachment to wealth; not the enjoyment of pleasurable things but the craving for them."[54] It is more important than ever to learn to distinguish between human needs and wants, between what is sufficient for human and planetary well-being and what constitutes craving and domination over nature and people.

The optimum balance between human well-being and environmental sustainability can be achieved by following a Middle Path toward the goods and services we all consume. One simple definition of poverty might be underconsumption, which contributes to human suffering. Excessive material consumption, on the other hand, also tends to decrease overall human well-being and leads to natural resource depletion, human alienation from nature, and spiritual emptiness.[55] Extreme economic inequalities also exacerbate fear, hatred, and social volatility. As the Middle Path Equilibrium Curve (Figure 5.1) illustrates, after a certain point, natural resource use and consumption brings diminishing returns socially and personally.

The call by some Western environmentalists for simpler living should not lead us to overlook the need for improved access to food, water, shelter, health care, and education for the poor. The world's resources are not unlimited. Increased consumption among the poor can only be achieved by reducing overconsumption by the wealthy. The exhortations of the Buddhist Middle Path, as of all the major religions, to cultivate generosity have to be translated into policies for more equitable and sustainable reallocation of global resources and the creation of livelihoods for all.

DOI: 10.1057/9781137308993

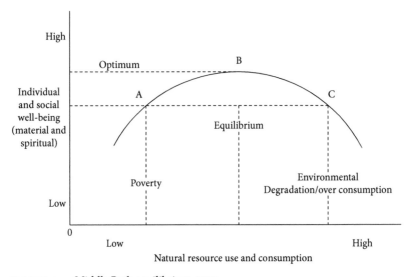

FIGURE 5.1   *Middle Path equilibrium curve*

Source: Adapted from Patrick Mendis, "Buddhist Equilibrium: The Theory of Middle Path for Sustainable Development," in Asoka Bandarage, *Women, Population and Global Crisis: A Political-Economic Analysis* (London: Zed Books, 1997), 331.

It is useful to consider the three broad classes of consumers identified by environmentalist Alan Durning and elaborated upon by economist David Korten: the over consumers, under consumers, and the sustainers.[56] About 80 percent of the world's environmental damage is attributed to the wealthiest 20 percent of the world's population—the over consumers who live mostly in the industrialized North and whose "lives are organized around [individually owned] cars, meat-based diets, use of packaged and disposable products."[57] At the bottom is the poorest 20 percent, who live predominantly in the Global South in "absolute deprivation," travelling mostly by foot, eating nutritionally inadequate diets, drinking contaminated water, using local biomass, and producing negligible wastes.[58] The overconsumption of the top rung and the underconsumption of the bottom are both unsustainable. The middle rung, the 60 percent also living mostly in the Global South—who travel mostly by bicycle and public surface transportation, eat healthy diets of grains, vegetables and some meat, use unpackaged goods, and recycle wastes—represent a balanced middle that should become the global norm.[59] Unfortunately, however, more and more countries are adopting the Western model of development based on overconsumption instead.

DOI: 10.1057/9781137308993

There is, however, an encouraging movement in the Global North toward moderation in the use of renewable sources of energy, organic agriculture, community gardening and marketing, cohousing, biking, public transportation, and recycling. These efforts need to be strengthened along with other efforts, such as cutting down meat consumption, which would contribute significantly to reducing carbon emissions as well as deforestation, soil erosion, water consumption, pesticide and fertilizer runoff, ground water depletion, and, of course, the subjection of animals to the terrible cruelty and suffering of unethical methods of production.[60] An April 2012 study from the Woods Hole Research Center estimates that the average person in the developed world would "need to cut meat consumption in half by the year 2050" to meet the emissions reduction targets set by the IPCC (Intergovernmental Panel on Climate Change).[61] As Rajendra Pachauri, the head of the IPCC, puts it, "In terms of immediacy of action and the feasibility of bringing about reductions in a short period of time, it [meat consumption] clearly is the most attractive opportunity."[62] A complete renunciation of meat eating is not required, only enjoyment in moderation with respect for the planet, animal life, and human health, including one's own health.

In the consumerist global culture of today, individuals have power as consumers. In addition to consuming less of unsustainably and unethically produced meat, people around the world can exert their consumer power through boycotts of genetically engineered food, violent entertainment, and a range of other harmful products and services. Brazilian scholar Leonardo Boff and other scholar-activists from around the world point out, that "the citizens' sense of responsibility as active agents of society" comes from participation in collective efforts.[63] The myriad local, national, and global single-issue movements need to be unified by a common vision of humanity as a species in nature.[64] Middle class intellectuals and professionals have a decisive role to play in this transformation. Do they identify with the dominant world order and the super elite or do they work toward environmental sustainability and the well-being of all?

To withstand external cooptation and repression, or internal dissension and rupture, social change movements must be grounded in a commitment to moral and ethical development and nonviolent change. The partnership society that we envision and our present ways of living need to be brought closer together through conscious individual and collective action. Philosopher Hannah Arendt said that "while strength is the natural quality of an individual seen in isolation, power springs

DOI: 10.1057/9781137308993

up between men when they act together and vanishes the moment they disperse."[65] The challenge, of course, is to mobilize this collective power for progressive ends—sustainability and the well-being of all—instead of extremist and sectarian purposes.

The modern world order holds within itself the basis of its own transformation. Modern science and technology allow people to see clearly the natural integration of the earth and humanity. Mass communication systems such as social media and the Internet do have the potential to help develop a global consciousness and interconnect social movements. To realize that potential, economic, and technological advancement has to be guided by ethical wisdom.[66] Sustainability and well-being cannot be postponed to the distant future. They must be upheld in how we think and live in the here and now. As the popular saying goes, "We are the ones we have been waiting for."[67]

Notes

1   Dieter Hessel, "Religion and Ethics Focused on Sustainability," in *Agenda for a Sustainable America*, ed. John C. Dernbach, 129 (Washington, DC: ELI Press, 2009).

2   "Ethics and Consciousness," www.mtnmath.com/implyh/node10.html (accessed Nov 21, 2012); www.worlddialogue.org; Gabriela Oliveira de Paula and Rachel Negrao Cavalcanti, "Ethics: Essence for Sustainability," *Journal of Cleaner Production* 8 (2000): 109–117; Roberto Savio, "Japan – Ethics, Democracy and Growth," Oct 2012; "Other News: Information that Markets Eliminate," www.other-news.info (accessed Nov 29, 2012).

3   Council for a Parliament of Religions, "Declaration Toward a Global Ethic," www.parliamentofreligions.org/_ . . . /TowardsAGlobalEthic.pdf.

4   Hessel, "Religion and Ethics," 132.

5   Earth Charter Initiative, www.earthcharterinaction.org/content/ . . . /read-the-charter.html; see also Global Greens Charter, www.globalgreens.org/globalcharter-english

6   "Einstein: Enigmatic Quote," icarus-falling.blogspot.com/2009/06/einstein-enigma.html (accessed Nov 28, 2012).

7   Daniel Goleman, *Emotional Intelligence* (New York: Bantam Books, 1995); Daniel Goleman, *Social Intelligence: The New Science of Human Relationships* (New York: Bantam Books, 2006).

8   Arthur Zajonc, "Cognitive-Affective Connections in Teaching and Learning: The Relationship Between Love and Knowledge," *Journal of Cognitive Affective Learning*, 3.1 (Fall 2006): 1.

DOI: 10.1057/9781137308993

 9 Martha E. Gimenez, "Revisiting the Marx-Malthus Debate," (Review of *Women, Population and Global Crisis*, by Asoka Bandarage), *Monthly Review*, 50.10 (March 1999): 51.

10 Riane Riane, *Chalice and the Blade: Our History, Our Future* (San Francisco: Harper and Row, 1987).

11 Janine M. Benyus, *Biomimicry: Innovation Inspired by Nature* (New York: Harper Perennial, 1997), 220, www.worlddialogue.org.

12 David Frawley, *Ayurveda and the Mind: The Healing of Consciousness* (Silver Lake, WI: Lotus Press, 1997); Rick Hanson, *Buddha's Brain: The Practical Neuroscience of Happiness, Love and Wisdom* (Oakland, CA: New Harbinger Publications, 2009).

13 Ronald L. Peters, *Exploring the Psychology of Disease: A Manual for Healing Beyond Diet and Fitness* (Nevada City: Blue Dolphin Publishing, 2003), 56; P. L. Dhar, "Dharma and Science," Vipassana Research Institute, www. vridhamma.org (accessed Nov 22, 2012); *The Yoga Sutras of Patanjali*, Translation and Commentary by Sri Swami Satchidananada (Yogaville: Integral Yoga Publications, 2007).

14 Cited in Peters, *Exploring the Psychology of Disease*, 58; see also Fritjof Capra, *The Tao of Physics* (Berkeley: Shambhala Publications, 1975).

15 For the teaching of S. N. Goenka on Vipassana Meditation, see www. dhamma.org (accessed Nov 26, 2012); Hanson, *Buddha's Brain*; "Investigating Healthy Minds with Richard Davidson," | On Being www.onbeing.org/ program/investigating-healthy-minds.../251date (accessed Nov 23, 2012).

16 M. Verhoeven, "Buddhism and Science: Probing the Boundaries of Faith and Reason," *Religion East and West*, 1 (June 2001): 77–97.

17 Charlene Spretnak, *The Resurgence of the Real: Body, Nature and Place in a Hypermodern World* (London: Routledge, 1999).

18 Eisler, *Chalice and Blade*.

19 Walpola Rahula, *What the Buddha Taught* (London: Gordon Fraser, 1978).

20 "Kohlberg's Stages of Moral Development," in W.C. Crain, *Theories of Development* (New York: Prentice-Hall, 1985), chapter 7, 118–136.

21 Felicia R. Lee, "Socially Engaged Without Preaching." *New York Times Online*, Oct 28, 2012 (accessed Nov. 21, 2012)

22 James Lovelock, *The Vanishing Face of Gaia: A Final Warning* (New York: Basic Books, 2010), 255.

23 Thich Nhat Hanh, *The Sun My Heart* (Berkeley: Parallax Press, 1988).

24 Gregory Bateson, *Steps to an Ecology of Mind: Collected Essays in Anthropology, Psychiatry, Evolution and Epistemology* (San Francisco: Chandler, 1972).

25 Cited in Bandarage, *Women, Population and Global Crisis*, 317.

26 "The Tsunami Tragedy: An Educational Forum," ISIM Publications, cdm15036.contentdm.oclc.org/cdm/singleitem/collection/.../11 (accessed Nov 28, 2012).

DOI: 10.1057/9781137308993

27   Al Gore, *Earth In the Balance: Ecology and the Human Spirit* (Boston: Houghton Mifflin, 1992), Conclusion.

28   See, for example, The Seville Statement of Violence of the 6th International Colloquium on Brain and Aggression, Seville, Spain, May 1986, cited in Bandarage, *Women, Population and Global Crisis*, Appendix 2, 348–350.

29   "Darwin the Buddhist? Empathy Writings Reveal Parallels," news. nationalgeographic.com/news/2009/02/090216-darwin . . . (accessed Nov 25, 2012); Amartya Sen, *Rationality and Freedom* (Cambridge: Harvard University Press, 2004).

30   Stanford University Center for Compassion and Altruism Research, ccare. stanford/education (accessed 23 Nov. 2012); Center for Investigating Healthy Minds, University of Wisconsin, Madison, www.onbeing.org/program/ investigating-healthy-minds . . . /251 (accessed Nov 23, 2012); California Institute for Integral Studies, www.ciis.edu (accessed Nov 22, 2012); Mind Life Institute, www.mindlife.org (accessed Nov 22, 2012).

31   Hans Kung, *A Global Ethic for Global Politics and Economics* (New York: Oxford University Press, 1998), 99.

32   His Holiness Tenzin Gyatso, The Fourteenth Dalai Lama, *A Human Approach to World Peace*, (London: Wisdom Publications, 1984), 4–5.

33   Bandarage, *Women, Population and Global Crisis*, 322.

34   Quoted in Verhoeven, "Buddhism and Science," 87; Philip Ryan, "Einstein's Quotes on Buddhism," Oct 26, 2007, Tricycle Editors' Blog. tricycleblog. wordpress.com/2007/10/26/einsteins-quotes-on-b . . . (accessed Feb 24, 2013)

35   Andrew Harvey, *The Hope: A Guide to Sacred Activism* (Carlsbad, CA: Hay House, 2009).

36   Ibid., 325; E. F. Schumacher, *Small is Beautiful* (New York: Harper and Row, 1989); "Towards A Sufficiency Economy: A New Ethical Paradigm for Sustainability," Jun 11, 2012, www.unesco.org/news . . . /towards-sufficiency-economy (accessed Nov 25, 2012).

37   R. Rockstrom et al., "Planetary Boundaries," *Ecology and Society*, 14.2, art. 32 (2009), www.ecologyandsociety.org.

38   Donella H. Meadows et al., *Limits to Growth: A Report for the Club of Rome's Project on the Predicament of Mankind* (New York: New American Library, 1977); M. Redclift, "Sustainable Development (1987–2005): An Oxymoron Comes of Age," *Sustainable Development*, 13.4 (2005): 212–227; Simon Dresner, *The Principles of Sustainability* (London: Earthscan, 2008), 29.

39   UN Commission on Environment and Development, *Our Common Future: Report of the World Commission on Environment and Development* (New York: Oxford University Press, 1987), 43; Herman E. Daly, *Beyond Growth: The Economics of Sustainable Development* (Boston: Beacon Press, 1996).

DOI: 10.1057/9781137308993

40   "The Six Nations: Oldest Living Participatory Democracy on Earth," www.
      ratical.org/many-worlds/6Nations (accessed Nov 23, 2012).

41   "A Core Principle," Women's Earth and Climate Caucus, Core Philosophy,
      www.iwecc.org/about-us/core-principle.php (accessed Nov 28, 2012).

42   Herman Daly, "Achieving Development Without Growth," *Surviving Together*
      (Winter 1993), 44.

43   Jeremy Rifkin, *The Third Industrial Revolution* (New York: Palgrave
      Macmillan, 2011).

44   Paul Hawken, Amory Lovins, and L. Hunter Lovins, *Natural Capitalism:
      Creating the Next Industrial Revolution* (Boston: Little Brown, 1999).

45   Benyus, "What Is Biomimicry?", www.asknature.org/article/view/what-is-
      biomimicry (accessed Nov 26, 2012).

46   Benyus, Biomimicry, 48–51.

47   "What is Biomimicry" www.ag.ndsu.nodak.edu/aginfo/ndipm/
      ipmdefinition.htm (accessed 23 Nov. 2012).

48   Bandarage, *Women, Population and Global Crisis*, 326.

49   Hazel Henderson, *Paradigms in Progress: Life Beyond Economics* (San
      Francisco: Berrett-Koehler Publishers, 1995).

50   Share the World's Resources, www.stwr.org; World Development Movement,
      www.wdm.org.uk; Friends of Earth, www.foe.org; Third World Network,
      www.twn.org.

51   Universal Declaration on Bioethics and Human Rights, http://portal.
      unesco.org/en/ev.php-URL_ID=31058&URL_DO=DO_TOPIC&URL_
      SECTION=201.html (accessed Nov 23, 2012); UN Declaration on Human
      Cloning, http://www.un.org/law/cloning/ (accessed Nov 23, 2012).

52   State of the World Forum, Statement on Life and Evolution, in Richard
      Heinberg, *Cloning the Buddha: The Moral Impact of Biotechnology* (New Delhi:
      Health Harmony, 1999), Appendix, 233–234.

53   Bill McKibben, *Enough: Staying Human in an Engineered Age* (New York:
      Henry Holt, 2003), 123; Martin Rees, *Our Final Century* (London: William
      Heinemann, 2003), chapter 6.

54   Schumacher, *Small is Beautiful*, 60.

55   Patrick Mendis, "Buddhist Equilibrium: The Theory of Middle Path for
      Sustainable Development," Staff Paper, 93–92, Department of Agricultural
      and Applied Economics, University of Minnesota, January 1993; Duane
      Elgin, *Voluntary Simplicity: Toward A Way of Life That is Outwardly Simple,
      Inwardly Rich* (New York: William Morrow, 1981).

56   Alan Thein Durning, *How Much is Enough: The Consumer Society and the
      Future of the Earth* (New York: W.W.Norton, 1992); Alan Durning, "Asking
      How Much Is Enough," *State of the World,* World Watch Institute, 1991; David
      C. Korten, *When Corporations Rule the World* (Bloomfield, Conn: Kumarian
      Press, 2001), 251–252.

DOI: 10.1057/9781137308993

57   Korten, When Corporations Rule, 251.

58   Ibid., 251–252.

59   Durning, "Asking How Much is Enough," 159.

60   "The Joy of Living Green," *Shambala Sun,* 20.2 (November 2011).

61   "To Reduce Global Warming, Address Meat Consumption," www. climatecentral.org/ . . . /to-reduce-global-warming (accessed Nov 2012); see also Scarcity vs. Distribution | A Well-Fed World. awellfedworld.org/issues/ scarcity

62   "UN Says Eat Less Meat to Curb Global Warming,"www.guardian.co.uk (accessed Nov 23, 2012).

63   Leonardo Boff, *Ecology and Liberation: A New Paradigm* (New York: Orbis Books, 1991), 83.

64   Asoka Bandarage, "In Search of a New World Order," *Women's Studies International Forum,* 14.4 (1991): 345–355.

65   Hannah Arendt, *The Human Condition* (Chicago: University of Chicago Press, 1958), 200.

66   Vaclav Havel, cited in Korten, *When Corporations Rule,* 337.

67   Alice Walker, *We Are the Ones We Have Been Waiting For* (New York: The New Press, 2006).

DOI: 10.1057/9781137308993

Bibliography

"10 FAQ on Biodiversity." *UNESCO.org*, UNESCO, n.d. Web. Nov 24, 2012.

350.org. 350.org Action Fund, n.d. Web. Nov 29, 2012.

"A Dictionary of Sociology." *Encyclopedia.com*. Web. Nov 20, 2012.

Abbott, Chris, Paul Rogers, and John Sloboda. *Briefing Paper June 2006, Global Responses to Global Threats*. Oxford: Oxford Research Group, 2006. Print.

Alcor Life Extension Foundation, "What Is Cryonics?" Web. *Cryonics*.

"Alter Nature: Designing Nature – Designing Human Life – Owning Life." *Z33: House for Contemporary Art Website*. Z33, Feb 18, 2011. Web. Nov 28, 2012.

"Atlas on Endangered Languages." *UNESCO.org*. UNESCO, Aug 23, 2004. Web. May 18, 2012.

Arendt, Hannah. *The Human Condition*. Chicago: U of Chicago P, 1958. Print.

Bacher, Dan. "CEOs of Big Foundations, Environmental NGOs Rake in the 'Green.'" *Daily Kos: News, Community, Action*. Kos Media, Aug 4, 2012. Web. Nov 28, 2012.

Bahro, Rudolf. *Avoiding Social and Ecological Disaster: The Politics of World Transformation*. Bath: Gateway, 1994. Print.

Bandarage, Asoka. "All Life Is Yoga." *Woman of Power* 4 (1986): 59, 83. Print.

_____. *Colonialism in Sri Lanka: The Political Economy of the Kandyan Highlands, 1833–1886*. Berlin: Mouton, 1983. Print.

_____. "Ethnic and Religious Tension in the World: A Political-Economic Perspective." *Global Political*

DOI: 10.1057/9781137308993

Economy and the Wealth of Nations: Performance, Problems and Future Prospects. Ed. Philip O'Hara. London: Routledge, 2004. Print.

"The 'Norwegian Model': Political Economy of NGO Peacemaking." *The Brown Journal of World Affairs* 17.2 (2011): 221–44. Print.

_____. "In Search of a New World Order." *Women's Studies International Forum* 14.4 (1991): 345–355. Print.

_____. *The Separatist Conflict in Sri Lanka: Terrorism, Ethnicity, Political Economy.* London: Routledge, 2009. Print.

_____. "Victims of Development." *Women's Review of Books* 5.1 (1987): 1, 3, 4. Print.

_____. *Women, Population and Global Crisis: A Political-Economic Analysis.* London: Zed Books, 1997. Print.

_____. 'Political Economy of Epidemic Kidney Disease in Sri Lanka' (forthcoming).

Barkley, Katherine, and Steve Weissman. "The Eco-Establishment." *Ramparts Magazine* (1970): 48–50. Print.

Barnett, Richard J. "The End of Jobs: Employment Is One Thing the Global Economy Is Not Creating." *Third World Resurgence* 44 (1994): 15–19. Print.

Barnosky, Anthony D., Matzke,Nicholas., Tomiya, Susumu., O.U. Wogan, Guinevere., Swartz, Brian., B.Tiago, Quental., McGuire, Jenny L., Lindsey, L. Emily, Maguire C. Kaitlin., Mersey, Ben., and Ferrer, A. Elizabeth "Has the Earth's Sixth Mass Extinction Already Arrived?" *Nature: International Weekly Journal of Science* 471(2011): 51–57. Web. Nov 23, 2012.

Baruah, Jamshed. "Snail's Pace towards Ban on Nuclear Testing." *IDN-In Depth News: Analysis That Matters.* IDN, Oct 2, 2012. Web. Nov 28, 2012.

Bateson, Gregory. *Steps to an Ecology of Mind: Collected Essays in Anthropology, Psychiatry, Evolution and Epistemology.* San Francisco: Chandler, 1972. Print.

Battiston, Stefano, Stefania Vitali, and James B. Glattfelder. "The Network of Global Corporate Control." *PLoS ONE* 6.10 (2011): n.p. Web. May 20, 2012.

Benyus, Janine M. *Biomimicry: Innovation Inspired By Nature.* New York: Harper Perennial, 1997. Print.

Biehl, Janet. "'Ecology' and the Modernization of Facism in the German Ultra-Right." *Ecofascism: Lessons from the German Experience.* By Biehl and Peter Staudenmaier. San Francisco: AK Press, 1995: 32–73. Print.

DOI: 10.1057/9781137308993

Bilchik, Gloria Shur. "Military Mystery: How Many Bases Does the US Have Anyway?" *Occasional Planet* Occasional Planet, Jan 24, 2011. Web. May 21, 2009.

Black Mesa Water Coalition Website. "Black Mesa Water Coalition," n.d. Web. Nov 28, 2012.

Block, Fred. "Reframing the Political Battle: Market Fundamentalism vs. Moral Economy." *Longview Institute: For the Public Good Website.* Longview Institute, n.d. Web. Nov 28, 2012.

Bodhi, Bhikku. "Tolerance and Diversity." *Buddhist Publication Society Newsletter* 24 (1993): 1–4. Print.

Boff, Leonardo. *Ecology and Liberation: A New Paradigm*, New York: Orbis, 1991. Print.

Bojanowski, Axel, and Christian Schwägerl. "The Anthropocene Debate: Do Humans Deserve Their Own Geological Era?" *Spiegel Online.* Spiegel Online International, Jul 8, 2011. Web. May 18, 2012.

_____. "UN Climate Body Struggling to Pinpoint Rising Sea Levels." *Speigel Online.* Speigel Online International, Jul 15, 2011. Web.

Bookchin, Murray. *The Ecology of Freedom: The Emergence and Dissolution of Hierarchy.* Oakland: AK Press, 2005. Print.

_____. *Our Synthetic Environment.* New York: Harper and Row, 1962. Print.

_____. *Post-Scarcity Anarchism.* Oakland: AK Press, 1971. Print.

Bostrom, Nick. "When Machines Outsmart Humans." *Futures* 35.7 (2000): 759–64. Print.

Boyce, Barry. "The Joy of Living Green." *Shambhala Sun.* 2011: 43–7. Print.

Budnik, Paul. "Ethics and Consciousness." *Mountain Math Software Website.* Web. Nov 21, 2012.

California Institute for Integral Studies Website. California Institute for Integral Studies, 2012. Web. Nov 22, 2012.

Capra, Fritjof. *The Tao of Physics.* Berkeley: Shambhala, 1975. Print.

Carson, Rachel. *Silent Spring*, Boston: Houghton Mifflin, 2002. Print.

Cavanaugh, John, and Frederic Clairmont. "The Rise of the TNC." *Third World Resurgence* 40 (1993): 19. Print.

The Center for Compassion and Altruism Research and Education Website. Stanford U, 2011. Web. Nov 23, 2012.

The Center for Investigating Healthy Minds Website. Waisman Center, U of Wisconsin-Madison, 2009. Web. Nov 28, 2012.

Centre for World Dialogue Website. Centre for World Dialogue, 2009. Web. Nov 28, 2012.

DOI: 10.1057/9781137308993

Chamie, Joseph. "Global Population of 10 Billion by 2100? – Not So Fast." *YaleGlobal Online.* Yale Center for the Study of Globalization, Oct. 26, 2011. Web.

Chavkin, Sasha. "As Kidney Disease Kills Thousands Across Continents, Scientists Scramble for Answers," The Center for Public Integrity, Oct 15, 2012. Web.

Cheeseman, Gina-Marie. "Oil Companies and Greenwashing." *Triple Pundit: People, Planet, Profit.* n.p. May 5, 2008. Web. Nov 28, 2012.

Chestney, Nina. "Global Warming Close to Becoming Irreversible." *Reuters Online* Mar 26, 2012. Web. Nov 29, 2012.

Chew, Kristina. "Dramatic Rise in 'Suicides By Economic Crisis' in Europe." *Care2 Website.* Care2.com, Inc, Apr 15, 2012. Web.

"Children's Rights." *BBC World Service Website.* BBC, n.d. Web. Nov 27, 2012.

Choi, Charles Q. "Human Marrying Robots? A Q&A with David Levy," *Scientific American,* Feb 14, 2008 Web. *Scientific.*

_____. "Forecast: Sex and Marriage with Robots," *Live Science* Oct 12, 2007. Web.

Chomsky, Noam. *New World of Indigenous Resistance.* Ed. Lois Meyer and Benjamin Maldonado. San Francisco: City Lights, 2010. Print.

Clark, John, Ed. Introduction. *Globalizing Civic Engagement: Civil Society and Transnational Action.* By Clark. London: Earthscan, 2003. Print.

Clarke, Chris. "SciAm Video Whiffs It on This Extinction Event." *Pharyngula: Evolution, Development, and Random Biological Ejaculations from a Godless Liberal.* n.p., Oct 21, 2012. Web. Nov 11, 2012.

Climate Refugees. (Dir) Michael Nash. LA Think Tank, 2010. Film.

Cochrane, Regina. "Rural Poverty and Impoverished Theory: Cultural Pluralism, Ecofeminism, and Global Justice." *Journal of Peasant Studies* 34.2 (2007), 167–206. Print.

"Convention on Biological Diversity: Farmers Demand an End to the Commercialization of Biodiveristy, GM Seeds and Synthetic Biology, Press Release." *La Via Campesina: International Peasant's Movement Website.* La Via Campesina, Oct 11, 2012. Web. Nov 28, 2012.

"A Core Principle." *IWECC.org.* Women's Earth and Climate Caucus, n.d. Web. Nov 23, 2012.

"Corporate Environmentalism: Enviro-sellouts." *Nonprofitwatch.org.* n.p., Nov 29, 2012. Web. Nov 29, 2012.

Coy, Peter. "The Youth Unemployment Bomb." *Bloomberg Businessweek Online* Feb 2, 2011: n.p. Web. Nov 29, 2012.

DOI: 10.1057/9781137308993

Crain, W. C. "Kohlberg's Stages of Moral Development." *Theories of Development.* By Crain. New York: Prentice Hall, 1985: 118–36. Print.

D'Almeida, Kanya. "U.S. Inaction on Climate Is 'Criminal', Activists Say." *Inter Press Service News Agency Website.* IPS, Dec 3, 2011. Web. Nov 29, 2012.

Daly, Herman E. "Achieving Development without Growth." *Surviving Together* (1993): 44–45. Print.

_____. *Beyond Growth: The Economics of Sustainable Development.* Boston: Beacon Press, 1996. Print.

"Declaration Toward a Global Ethic." *Parliament of the World's Religions Website.* Council for a Parliament of the World's Religions, Sep 4, 1993. Web. Nov 27, 2012.

"Definitions." *University of Redding ECIFM Website.* University of Redding, n.d. Web. Nov 23, 2012.

Dell'Amore, Christine. "Darwin the Buddhist? Empathy Writings Reveal Parallels." *National Geographic Online.* National Geographic News, Feb 16, 2009. Web. Nov 28, 2012.

Dhar, P. L. "Dharma and Science." *Vipassana Research Institute Website.* Vipassana Research Institute, 2010. Web. Nov 22, 2012.

Diamond, Jared. *Collapse: How Societies Choose to Fail or Succeed.* New York: Viking Press, 2005. Print.

Diener, E. "Subjective Well-being. The Science of Happiness and a Proposal for a National Index." *The American Psychologist* 55.1 (January 2000): 34–43. Print.

"Director: Richard J. Davidson, PhD." *The Center for Investigating Healthy Minds Website.* Waisman Center, U of Wisconsin-Madison, 2009. Web. Nov 28, 2012.

Dresner, Simon. *The Principles of Sustainability.* London: Earthscan, 2008. Print.

Drum, Kevin. "Chart of the Day: Our Robot Overlords Will Take Over Soon." *Mother Jones Online.* Mother Jones and the Foundation for National Progress, Apr 17, 2012. Web. Nov 12, 2012.

Durning, Alan T. *How Much Is Enough: The Consumer Society and the Future of the Earth.* New York: Norton, 1992. Print.

"The Earth Charter." *The Earth Charter Initiative Website.* Earth Charter Initiative, 2012. Web. Nov 29, 2012.

"The Economic and Politics of the World Social Forum: Lessons for the Struggles Against 'Globalisation.'" *Global Research Online.* Global Research, 2004. Web. Nov 28, 2012.

DOI: 10.1057/9781137308993

Eisler, Riane. *The Chalice and the Blade: Our History, Our Future.* San Francisco: Harper and Row, 1987. Print.

Elgin, Duane. *Voluntary Simplicity: Toward A Way of Life That Is Outwardly Simple, Inwardly Rich.* New York: William Morrow, 1981. Print.

"Extinction Crisis Continues Apace." *International Union for Conservation of Nature Website.* IUCN, Mar 2009. Web. Nov 28, 2012.

"Facts about the Rainforest." *Save the Rainforest Website.* Save the Rainforest, n.d. Web. May 18, 2012.

Fahey, David W., et al. Twenty Questions and Answers about the Ozone Layer: 2010 Update. DC: NOAA, 2010. Print.

"FAO Accused of 'Promoting the Destruction of Peasant and Family Farming'." *Social Watch: Poverty Eradication and Gender Justice.* Social Watch News, Sep 20, 2012. Web. Nov 28, 2012.

Farrar, Lara. "Scientists: Humans and Machines Will Merge in Future." *CNN.com/technology.* CNN, Jul 15, 2008. Web. Nov 28, 2012.

Feshbach, Murray. *Ecocide in the USSR: Health and Nature Under Siege.* New York: BasicBooks, 1992. Print.

Feuerstein, Georg, and Brenda Feuerstein. *Green Yoga.* Saskatchewan: Traditional Yoga Studies, 2007. Print.

Fidelity Investments. "Synthetic Biology – Fidelity Investments: Thinking Big." *YouTube.* Web. Nov 28, 2012. www.youtube.com/playlist?list=FLILvMv--Jgq...

Fields, Gary. *Poverty, Inequality and Development.* Cambridge: Cambridge University Press, 1980. Print.

Fincher, Lindsay. "Former Soviet Republics Top List of 'World's Most Polluted Places'." *At Home In The Wasteland Travel Blog.* WordPress, n.d. Web. Nov 28, 2012.

Frawley, David. *Ayurveda and the Mind: The Healing of Consciousness.* Silver Lake: Lotus, 1997. Print.

Freeland, Chrystia. *Plutocrats: The Rise of the New Global Super-Rich and the Fall of Everyone Else.* New York: Penguin Press, 2012. Print.

Freedman, Andrew. "To Reduce Global Warming, Address Meat Consumption." *Climate Central.* Climate Central, Apr 13, 2012. Web. Nov 12, 2012.

Friends of the Earth. "Nanotechnology, Climate and Energy: Over-Heated Promises and Hot Air?" *Friends of the Earth* 2010. Web. Nov 28, 2012.

Friends of the Earth Website. Friends of the Earth U.S., n.d. Web. Nov 28, 2012.

DOI: 10.1057/9781137308993

Frey, Bruno, and Alois Stutzer. *Happiness and Economics.* Princeton: Princeton UP, 2002. Print.

Friedman, Lisa. "Climate Change Makes Refugees in Bangladesh." *Scientific American.* Scientific American, Inc., Mar 3, 2009. Web. May 18, 2012.

Friedman, Milton. *Capitalism and Freedom.* Chicago: U of Chicago P, 1962. Print.

Friedman, Thomas. "Made in the World." *The New York Times* Jan 29, 2012: n.p. Print.

Fukuda-Parr, Sakiko and Shiva Kumar A.K. *Readings in Human Development: Concepts, Measures and Policies for a Development Paradigm,* Oxford: Oxford University Press, 2005. Print.

Fukuyama, Francis. *Our Posthuman Future: Consequences of the Biotechnology Revolution.* New York: Picador, 2002. Print.

Georgetown University, "The Tsunami Tragedy: An Educational Forum." Transcript of proceedings. Georgetown U Institute for the Study of International Migration, 2005. Web. Nov 28, 2012.

Giddings, Bob, Bill Hopwood, and Geoff O'Brien. "Environment, Economy and Society: Fitting Them Together into Sustainable Development." *Sustainable Development* 10.4 (2002): 187–96. Print.

Gies, Erica. "Holding on to What Was in the Andamans." *New York Times* Feb 12, 2012 Sunday Travel Section: 12. Print.

Gilbertson, Tamra, and Oscar Reyes. "Carbon Trading – How It Works and Why It Fails." *Critical Currents: Dag Hammerskjöld Foundation Occasional Paper Series* 7 (2009): pp. 7–104. Web. May 17, 2012.

Gimenez, Martha E. "Revisiting the Marx-Malthus Debate." Review of *Women, Population and Global Crisis,* by Asoka Bandarage. *Monthly Review* 50.10 (March 1999): n.p. Print.

"Global: UN- Over Two Billion People Now Connected to Internet But Digital Divide Remains Wide." *UNPAN Global.* UN Public Administration Network, Nov 13, 2012. Web. Nov 28, 2012.

Global Change Master Directory Website. NASA: Goddard Space Flight Center, Nov 2012. Web. Nov 28, 2012.

"Global Greens Charter." *Global Greens Website.* Global Greens Network, 2012. Web. Nov 29, 2012.

"Global HIV and AIDS Estimates, 2009 and 2010." *Worldwide HIV & AIDS Statistics.* AVERT. Web. Nov 28, 2012.

"Global Meat Production and Consumption Continue to Rise." *WorldWatch Website.* WorldWatch Institute, n.d. Web. May 16, 2012.

DOI: 10.1057/9781137308993

"Global Methane Initiative." *US Environmental Protection Agency*, n.d. Web. May 16, 2012.

"Global Youth Unemployment at All-Time High: UN." *Daily Star Online.* Aug 13, 2011: n.p. Web. Nov 28, 2012.

Goenka, S. N. "Vipassana Meditations." *Dhamma.org.* n.p., Aug 22, 2012. Nov 28, 2012.

"Going beyond GDP, UNDP Proposes Human Development Measure of Sustainability." *World Development Reports Website.* United Nations Development Programme, Jun 20, 2012. Web.

Goldin, Ian and Reinert, Kenneth A. *Globalization for Development: Meeting New Challenges.* New York: Oxford University Press, 2012. Print.

Goldsmith, Edward. "My Answer." *Edward Goldsmith: The Late Environmentalist, Author, & Philosopher.* Website. The Estate of Edward Goldsmith, Jan 1, 2003. Web.

Goleman, Daniel. *Emotional Intelligence: Why It Can Matter More than IQ.* New York: Bantam Books, 1995. Print.

_____. *Social Intelligence: The New Science of Human Relationships.* New York: Bantam Books, 2006.

Gore, Albert, Jr. *Earth in the Balance: Ecology and the Human Spirit.* Boston: Houghton Mifflin, 1992. Print.

Groundswell International Website. Groundswell International, n.d. Web. Nov 28, 2012.

Gunder, Michael. "Sustainability: Planning's Redemption or Curse?" *Planetizen.* Urban Insight, Inc, Feb 8, 2007. Web. Nov 29, 2012.

"Half of Mammals 'in Decline' Says Extinction Red List." *Associated Foreign Press.* Google News, Oct 6, 2008. Web. Nov 29, 2012.

Halperin, James L. *The First Immortal.* New York: Ballantine, 1998. Print.

Hamilton, Clive. *Growth Fetish.* London: Pluto, 2003. Print.

Hanh, Thich Nhat. *The Sun My Heart.* Berkeley: Parallax, 1988. Print.

Hansen, James. *Storms of My Grandchildren: The Truth About the Coming Climate Catastrophe and Our Last Chance to Save Humanity*, New York: Bloomsbury, 2009.

_____."Air Pollutant Climate Forcings Within the Big Climate Picture." *Climate Change Congress, "Global Risks, Challenges & Decisions"*, Copenhagen, Mar 11, 2009. n.p. Web.

Hanson, Rick. *Buddha's Brain: The Practical Neuroscience of Happiness, Love and Wisdom.* Oakland: New Harbinger, 2009. Print.

Happy Planet Index Website. New Economics Foundation, n.d. Web. Sep 11, 2012.

DOI: 10.1057/9781137308993

Hari, Johann. "The Wrong Kind of Green." *Nation* Mar 4, 2010: n.p. Web. Nov 23, 2012.

Harris, Marvin and Eric B. Ross. *Death, Sex and Fertility: Population Regulation in Preindustrial and Developing Societies.* New York: Columbia UP, 1987. Print.

Harvey, Andrew. *The Hope: A Guide to Sacred Activism.* Carlsbad, CA: Hay House, 2009

Hawken, Paul, Amory Lovins, and L. Hunter Lovins. *Natural Capitalism: Creating the Next Industrial Revolution.* Boston: Little Brown, 1999. Print.

Hawkins, William R. "Green Terrorism's New Left Roots." *Frontpagemag. com.* Sep 8, 2010: n.p. Web. Nov 28, 2012.

Heinberg, Richard. *Cloning the Buddha: The Moral Impact of Biotechnology.* New Delhi: Health Harmony, 1999. Print.

Hedges, Chris. *Days of Destruction: Days of Revolt.* New York: Nation, 2012 Print.

Henderson, Hazel. *Paradigms in Progress: Life Beyond Economics.* San Francisco, Berrett-Koehler, 1995. Print.

Hessel, Dieter. "Religion and Ethics Focused on Sustainability." *Agenda for a Sustainable America.* Ed. John C. Dernbach. Washington, DC: Environmental Law Institute, 2009. 129–41. Print.

Heuvel, Katrina. "Ending War for Profit." *Share the World's Resources: Sustainable Economics to End Global Poverty.* Share the World's Resources, Sep 29, 2007. Web. Nov 28, 2012.

Hildyard, Nicholas. "'Blood' and 'Culture': Ethnic Conflict and the Authoritarian Right." *Corner House Briefing Papers* (No. 11). Corner House, Jan 29, 1999. Web. Nov 27, 2012.

Hofstadter, Richard. *Social Darwinism in American Thought.* New York: George Braziller, 1959. Print.

"How Will Global Warming Change Earth?" *Earth Observatory.* NASA, n.d. Web. May 18, 2012.

Howarth, Robert W. and Alan B. Townsend. "Fixing the Global Nitrogen Problem." *Scientific American Online.* Scientific American, Inc, Jan 27, 2010. Web. Nov 29, 2012.

Hribar, Carrie. *Understanding Concentrated Animal Feeding Operations and Their Impact on Communities.* Bowling Green: National Association of Local Boards of Health, 2010. Print.

Hrushka, Bronwen. "Raising the Ritalin Generation." *New York Times* Aug 18, 2012, weekend ed. SR6. Print.

DOI: 10.1057/9781137308993

"Human Nature on Collision Course with Genetic Engineering."
WorldWatch Institute Website. World Watch Institute, 2012. Web.
Nov 29, 2012.

"Hunger Stats." World Food Programme: Fighting Hunger Worldwide.
WFP, n.d. Web. Nov 29, 2012.

Huxley, Aldous. *Brave New World.* New York: HarperCollins, 1998. Print.

Institute for American Values, 'Hard Wired to Connect the New
Scientific Case for Authoritative Communities, 2003. Print.

International Council for Local Environmental Initiatives (ICLEI)
Local Agenda 21 Initiative. *The Local Agenda 21 Planning Guide: An
Introduction to Sustainable Development Planning (electronic edition),*
1996. Web. Nov 28, 2012.

International Organization for Migration Website. International
Organization for Migration, n.d. Web. Nov 28, 2012.

"Interview with Ted Kaczynski, Administrative Maximum Facility
Prison, Florence, Colorado, USA." *Earth First Journal* June 1999. Web.
Nov 28, 2012.

"Iraq Deaths." *Just Foreign Policy.* n.d. Web. May 21, 2012.

Irfan, Umair. "Nitrogen Pollution Likely to Increase Under Climate
Change." *Scientific American Online.* 16 2012: n.p. Web. Nov 28, 2012.

"IUCN Cries Foul Over Trade in Python Skins But CITES Issued
400,000 Export Licenses." *WildlifeExtra.com.* Wildlife Extra, n.d.
Web. Nov 28, 2012.

Jowit, Juliette. "UN Says Eat Less Meat to Curb Global Warming."
Observer Online. Sep 6, 2008: n.p. Web. Nov 23, 2012.

Joy, Bill. "Why the Future Does Not Need Us." *Wired* 8.04 (2000): n.p.
Print.

Kamal, Sajed. *The Renewable Revolution: How We Can Fight Climate
Change, Prevent Energy Wars, Revitalize the Economy and Transition
to a Sustainable Future.* London: Earthscan, 2011. Print.

Khor, Martin. "Climate Talks at New Crossroads." *The Star Online* Sep 17,
2012. Web. Nov 23, 2012.

_____. "World Wide Unemployment Will Reach Crisis Proportions Says
Social Expert." *Third World Resurgence* 44 (1994): n.p. Print.

Klare, Michael T. *Resource Wars: The New Landscape of Global Conflict.*
New York: Henry Holt, 2001. Print.

Klein, Naomi. *Shock Doctrine.* New York: Henry Holt, 2008. Print.

Knight, Matthew. "U.N. Report: Eco-Systems at 'Tipping Point.'"
CNNWorld Online. CNN, May 10, 2011: n.p. Web. Nov 28, 2012.

DOI: 10.1057/9781137308993

Korten, David. *When Corporations Rule the World*. Bloomfield: Kumarian, 2001. Print.

Kung, Hans. *A Global Ethic for Global Politics and Economics*. New York: Oxford UP, 1998. Print.

Kurzweil, Ray. *The Age of Spiritual Machines: When Computers Exceed Human Intelligence*. Middlesex: Penguin, 2000. Print.

Lasch, Christopher. *Culture of Narcissism: American Life in An Age of Diminishing Expectations*. New York: Norton, 1979. Print.

Lawrence, Felicity. "Food Prices to Double by 2030, Oxfam Warns." *The Guardian Online*. Guardian News, May 31, 2011. Web. May 20, 2012.

_____."The Global Food Crisis: ABCD of Food – How the Multinationals Dominate Trade." *Poverty Matters Blog*. Guardian News, Jun 2, 2011. Web. May 20, 2012.

Layard, Richard. *Happiness: Lessons From a New Science*. New York: Penguin, 2005. Print.

Lee, Felicia R. "Socially Engaged Without Preaching." *New York Times Online* Oct 28, 2012. Web. Nov 21, 2012.

Lerner, Gerda *The Creation of Patriarchy*, Oxford: Oxford University Press, 1986.

Li, Huey-li. "A Cross-Cultural Critique of Ecofeminism." *Ecofeminism: Women, Animals, Nature*. Ed. Greta Gaard. Philadelphia: Temple UP, 1993. Print.

Lorimer, David. *Radical Prince: The Practical Vision of the Prince of Wales*. Edinburgh: Floris, 2003. Print.

Lovelock, James. *The Vanishing Face of Gaia: A Final Warning*. New York: Basic Books, 2010. Print.

MacDonald, Christine. *Green Inc: An Environmental Insider Reveals How a Good Cause Has Gone Bad*. Connecticut: Lyons, 2008. Print.

McCoy, Alfred W. "Space Warfare and the Future of US Global Power," *Mother Jones*, Nov 8, 2012, Web.

"The Making of a World Water Crisis." *Stopcorporateabuse.org*. Corporate Accountability International, n.d. Web. May 21, 2012.

Makwana, Rajesh. "Multinational Corporations (MNCs): Beyond the Profit Motive." *Share The World's Resources*. Share the World's Resources, Oct 3, 2006. Web. Nov 29, 2012.

Malone, Andrew. "The gm Genocide: Thousands of Indian Farmers Are Committing Suicide after Using Genetically Modified Crops." *Mail Online*. The Guardian, Nov 2, 2008. Web. May 21, 2012.

DOI: 10.1057/9781137308993

Manser, Ann. "Out of Africa: Anthropologist Who Discovered 'Lucy' Discusses Human Origins." *UDaily.* University of Delaware, Apr 22, 2011. Web. Nov 28, 2012.

Marx, Karl. *The Future Results of the British Rule in India, in On Colonialism: Articles from the New York Tribune and Other Writings.* Ed. Frederick Engels and Karl Marx. New York: International Publishers, 1972. Print.

McCarthy, Michael. "Oceans on Brink of Catastrophe." *Independent Online.* Jun 21, 2011: n.p. Web. Nov 27, 2012.

McKibben, Bill. *Enough: Staying Human in an Engineered Age.* New York: Henry Holt, 2003. Print.

_____. "Human Flourishing Depends on What We Do Now." *Ecologies of Human Flourishing.* Ed Donald K. Swearer and Susan Lloyd McGarry. Cambridge: Harvard UP, 2011. Print.

_____. *Nanotechnology, Climate and Energy: Over-Heated Promises and Hot Air?* UK: Friends of Earth International, 2010. Print.

Mendis, Patrick. "Buddhist Equilibrium: The Theory of Middle Path for Sustainable Development." Staff Paper P93–2. University of Minnesota, 1993. Print.

Merchant, Carolyn. *Death of Nature: Women, Ecology and the Scientific Revolution.* San Fransisco: Harper and Row, 1983. Print.

Messenger, Stephen. "Climate Refugees Could Number 1 Billion by 2050." *Treehugger.* MNN Holdings LLC, Dec 8, 2009. Web. May 17, 2010.

Mick, Jason. "Conference Examines Man and Machine Merging, How Tech Will Make Human Brain Obsolete." *DailyTech Online.* DailyTech, Jul 15, 2008. Web. Nov 28, 2012.

"Military Expenditure." Stockholm International Peace Research Institute. SIPRI. Web. Nov 28, 2012.

Mind & Life Institute Website. Mind & Life Institute, n.d. Web. Nov 29, 2012.

Mishel, Lawrence. "Education Is Not the Cure for High Unemployment or for Income Inequality," Jan 12, 2011. Economic Policy Institute, Web.

Monbiot, George. "Black Shirts in Green Trousers." *Guardian*, Apr 30, 2002: n.p. Web. Nov 28, 2012.

Mumford, Lewis. *Technics and Civilization*, New York: Harcourt, 1934.

Monsanto Company. "Farmer Suicides in India – Is There a Connection with Bt Cotton?" *Monsanto.com.* Monsanto Company, n.d. Web. Nov 28, 2012.

"NASA Study Finds World Warmth Edging Ancient Levels." *NASA: Goddard Institute for Space Studies*, Sep 25, 2006. Web. May 16, 2012.

DOI: 10.1057/9781137308993

Niles, Chris. "Meeting Notes Progress on Reducing Maternal and Child Mortality." *UNICEF Website.* UNICEF, Sep 24, 2012. Web. Nov 28, 2012.

"NOAA: Past Decade Warmest on Record According to Scientists in 48 Countries." *NOAA Website.* NOAA, Jul 28, 2010. Web. Nov 28, 2012.

Norberg-Hodge, Helena. *Ancient Futures: Learning from Ladakh for a Globalizing World.* San Fransisco: Sierra Club, 2009. Print.

Nordland, Rod. "Risks of Afghan War Shift from Soldiers to Contractors." *New York Times.* Feb 11, 2011: A1. Print.

Norrell, Brenda. "Indigenous March on Wednesday to Deliver Kari-Oca Declaration to World Leaders in Rio." *Indigenous Environmental Network – Road to Rio+20 and Beyond.* Indigenous Environmental Network, n.d. Web. Oct 7, 2012.

"OECD Environmental Outlook to 2050: The Consequences of Inaction." *OECD iLibrary.* OECD, Mar 15, 2012. Web.

Oliveira de Paula, Gabriela, and Rachel Negrao Cavalcanti, "Ethics: Essence for Sustainability." *Journal of Cleaner Production* 8 (2000): 109–117. Print.

Olivera, Oscar, and Tom Lewis. *¡Cochabamba!: Water War in Bolivia.* Cambridge: South End, 2008. Print.

Orwell, George. *Nineteen Eighty Four,* New York: New American Library, 1949.

Pandaram, Jamie. "Diabetes Threatens Aborigine Extinction." *The Sydney Morning Herald Online.* The Sydney Morning Herald, Nov 14, 2006. Web.

"Paradigm." The American Heritage® Dictionary of the English Language. Boston: Houghton Mifflin, 2000. TheFreeDictionary.com. Web. Nov 12, 2012.

Patel, Raj. *The Value of Nothing: How to Reshape Market Society and Redefine Democracy.* New York: Picador, 2009. Print.

_____. The Long Green Revolution," *The Journal of Peasant Studies* 40:1 (2013), pp. 1–63. Print.

Paul, Richard, and Linda Elder. *The Miniature Guide to Critical Thinking-Concepts and Tools.* Tomales, CA: The Foundation for Critical Thinking, 2003. Print.

Peter, Sunny. "India: People's Movement Marches to Victory." *Lapidomedia: Centre for Religion in World Affairs Website.* Lapido Media, Oct 15, 2012. Web. Nov 26, 2012.

Peters, Ronald L. *Edgework: Exploring the Psychology of Disease: A Manual for Healing Beyond Diet and Fitness.* Nevada City: Blue Dolphin, 2003. Print.

DOI: 10.1057/9781137308993

"Pioneering Study Shows Richest Two Percent Own Half World Wealth."
    UNU-Wider. United Nations University, Dec 5, 2006. Web. Nov 28, 2012.
Pizzigati, Sam. "The Difference More Global Equality Could Make." *Too
    Much Online.* Program on Inequality and the Common Good at the
    Institute for Policy Studies, Oct 17, 2012. Web. Nov 12, 2012.
Polaris Institute, 'United Nations Indicted for Enabling Corporate
    Control of Water and Greenwashing,' Jan 27, 2010. Web. May 21, 2010.
Polyani, Karl. *The Great Transformation.* Boston: Beacon Press, 1957. Print.
PopDev: Think. Act. Connect. Website. Population and Development
    Program, based at Hampshire College, 2008. Web. Nov 29, 2012.
Porritt, Jonathon. *Capitalism as If the World Matters.* London: Earthscan,
    2005. Print.
Postman, Neil. *Amusing Ourselves to Death: Public Discourse in the Age of
    Show Business.* New York: Penguin Books, 1985. Print.
"Press Release: Organizations Renew Demand to UN for Global Ban
    on GM Trees." *Global Justice Equality Project Website.* Global Justice
    Equality Project, Oct 2012. Web. Nov 28, 2012.
Project Censored: Media Democracy in Action Website. n.p., n.d. Web.
    Nov 23, 2012.
"Rainforest Destruction." *AG101.* CSU-Pomona. Web. May 21, 2012.
Rahula, Walpola. *What the Buddha Taught.* London: Gordon Fraser, 1978.
    Print.
"Rates of Rainforest Destruction and Species Loss in the Ecuadorian
    Amazon." *EcuadorExplorer.com.* Rainforest Action Network. Web.
    Nov 28, 2012.
Rees, Martin. *Our Final Century.* London: William Heinemann, 2003. Print.
Redclift, Michael. "Sustainable Development (1987–2005): An
    Oxymoron comes of Age." *Sustainable Development* 13.4 (2005):
    212–27. Print.
Rifkin, Jeremy. *The End of Work: The Decline of the Global Labor Force and
    the Dawn of the Post-Market Era.* New York: G. P. Putnam's Sons, 1995.
    Print.
_____. "New Technology and the End of Jobs." *Converge.* n.p., n.d. Web.
_____. *The Third Industrial Revolution: How Lateral Power Is Transforming
    Energy, the Economy, and the World.* New York: Palgrave Macmillan,
    2011. Print.
"Rise in International Arms Transfers Is Driven by Asian Demand, Says
    SIPRI." *Stockholm International Peace Research Institute.* SIPRI, Mar 19,
    2012. Web.

DOI: 10.1057/9781137308993

Rockström, J., Steffen W., Noone, K., Persson, A., Chapin, F.S. III., Lambin, E., Lenton,T.M., Scheffer, M., Folke, C., Schellnhuber, H., Nykvist, B., De Wit, C.A., Hughes, T., van der Leeuw, S., Rodhe, H., Sörlin, S., Snyder, P.K., Costanza, R., Svedin, J., Falkenmark, M., Karlberg, L., Corell, R.W., Fabry, V.J., Hansen, J., Walker, B., Liverman, D., Richardson, K., Crutzen,P., Foley, J. "Planetary Boundaries :Exploring the Safe Operating Space for Humanity." *Ecology and Society* 14.2 (2009): n.p. Web.

Rohr, Bob. "AAAS Coalition Explores Perspectives of Indigenous Communities on Climate Change." *American Association for the Advancement of Science*, Feb 6, 2012. Web. May 18, 2012.

Rouse, Jeremy David, et al. "Nitrogen Pollution: An Assessment of Its Threat to Amphibian Survival." *Environment Health Perspectives* 107.10 (1999): 799–803. Print.

Rubenstein, Madeleine. "A Changing Climate for Small Island States." *State of the Planet: Blogs from the Earth Institute.* The Earth Institute at Columbia U, Dec 15, 2011 Web. May 18, 2012.

Philip Ryan, "Einstein's Quotes on Buddhism," *Tricycle, Einstein's Quotes* on *Buddhism,* Oct. 26,2007, Tricycle Editors' Blog Oct 26, 2007 Web.

Salmon, Felix. "The Global Youth Unemployment Crisis." *Reuters Online.* Dec 22, 2011: n.p. Web. Nov 27, 2012.

Satchidananda, Sri Swami. *The Yoga Sutras of Patanjali.* Buckingham: Integral Yoga Publications, 1990. Print.

Savio, Roberto. "Japan – Ethics, Democracy, and Growth." *Other News: Information that markets eliminate.* Other News, Oct 22, 2012. Web. Nov 28, 2012.

"Scarcity vs. Distribution." *A Well-Fed World Website.* A Well-Fed World., n.d. Web. Nov 2012.

Schimmel, Annemarie *Mystical Dimensions of Islam,* Chapel Hill: University of North Carolina Press, 2011. Print.

Schönleitner, Günther. "World Social Forum: Making Another World Possible?" *Globalizing Civic Engagement: Civil Society and Transnational Action.* Ed. John Clark. London: Earthscan, 2003: 127–49. Print.

Schumacher, E. F. *Small Is Beautiful: Economics as If People Mattered.* New York: Harper and Row, 1973. Print.

Scitovsky, Tibor. *The Joyless Economy: The Psychology of Human Satisfaction,* Oxford: Oxford UP, 1992.

Scott, James C. *Weapons of the Weak: Everyday Forms of Peasant Resistance.* New Haven: Yale UP, 1985. Print.

DOI: 10.1057/9781137308993

Seed, John, et al. *Thinking Like A Mountain.* Philadelphia: New Society, 1988. Print.

Sen, Amartya. *Development As Freedom.* New York: Knopf, 2000. Print.

_____. "Human Development Index" in *The Elgar Companion To Development Studies.* Ed. David Alexander Clark. Cheltenham, UK: Edward Elgar Publishing, 2007 Print.

_____. *Rationality and Freedom,* Cambridge: Harvard University Press, 2004. Print.

Shapiro, Judith. *Mao's War against Nature.* New York: Cambridge UP, 2001. Print.

Share the World's Resources Website. Share the World's Resources, n.d. Web. Nov 28, 2012.

Shaw, Anny. "Last Member of 65,000-Year-Old Tribe Dies, Taking One of World's Earliest Languages to The Grave." *Mail Online.* Associated Newspapers Ltd, Feb 9, 2010. Web.

Shiva, Vandana. *Staying Alive.* Boston: New Society, 1988. Print.

Shiva, Vandana and Maria Mies. *Ecofeminism.* London: Zed, 1993. Print.

Shostack, Marjorie. *Nisa: The Life and Words of a !Kung Woman.* New York: Vintage, 2000. Print.

Shreeve, Jamie. "Oldest Skeleton of Human Ancestor Found." *National Geographic News Online* Oct 1, 2009: n.p. Web. Nov 28, 2012.

Singe, Jaysen, Ed. *Becoming Human Website.* Institution of Human Origins, 2000. Web. Nov 27, 2012.

"The Six Nations: Oldest Living Participatory Democracy on Earth." *Ratical.org.* Six Nations Indian Museum, n.d. Web. Nov 23, 2012.

Smith, Adam. *An Inquiry into the Nature and Causes of the Wealth of Nations.* London: Methuen, 1904. Print.

Spretnak, Charlene. *The Resurgence of the Real: Body, Nature and Place in a Hypermodern World.* London: Routledge, 1999. Print.

Staudenmaier, Peter. " Fascist Ecology: The 'Green Wing' of the Nazi Party and Its Historical Antecedents." *Ecofascism: Lessons from the German Experience.* By Janet Biehl and Peter Staudenmaeir. San Francisco: AK Press, 1995. Print.

Stea, Carla. "Manipulation of the UN Security Council in Support of the US-NATO Military Agenda." *Global Research Online.* Center for Research on Globalization, Jan 10, 2012. Web. May 19, 2012.

Steady, Filomina Chioma, ed. *Environmental Justice in the New Millennium: Global Perspectives on Race, Ethnicity, and Human Rights.* New York: Palgrave Macmillan, 2009. Print.

DOI: 10.1057/9781137308993

Stiglitz, Joseph. *Making Globalization Work.* New York: Norton, 2006. Print.

Stiglitz, Joseph, Amartya Sen, and Jean-Paul Fitoussi. "Report of the Commission on the Measurement of Economic Performance and Social Progress." *Commission on the Measurement of Economic Performance and Social Progress,* Sep 14, 2009. Print.

Strange, Susan. *Casino Capitalism.* Manchester: Manchester UP, 1997. Print.

Strieker, Gary. "Scientists Agree World Faces Mass Extinction." *CNN. com/Sci-Tech.* CNN, Aug 2002. Web. Nov 23, 2012.

Stusser, Michael A. "The Tweeting, Yelping, Flickring, Foursquaring, TripAdvising Mentality." *Shambhala Sun.* Sep 27, 2012. Print.

Summers, Lawrence H. "World Bank Office Memorandum, 12 Dec 1991." Reprinted by *The Economist* Feb 8, 1992: 66. Print.

Survival International, "Two Years After Andaman Tribe Dies, Another 'Faces Extinction,'" Jan 26, 2012. Web. Nov 28, 2012.

"The Sustainability Challenge." *The Sustainability Laboratories Website.* The Sustainability Laboratories, 2009. Web. Nov 29, 2012.

Swimme, Brian Thomas and Mary Evelyn Tucker. *Journey of the Universe.* New Haven: Yale UP, 2011. Print.

Third World Newsletter Website. Third World News, n.d. Web. Nov 29, 2012.

Thompson, E. P. *The Making of the English Working Class.* New York: Penguin, 1991. Print.

_____."Notes on Exterminism: The Last Stage of Civilization." *New Left Review* 121 (1980): 3–31.

_____. "Time, Work-Discipline and Industrial Capitalism." *Past and Present* 38 (1967): 56–97. Print.

Tilt, Bryan. *The Struggle for Sustainability in Rural China.* New York: Columbia UP, 2010. Print.

"Towards A Sufficiency Economy: A New Ethical Paradigm for Sustainability." *Conference Proceedings.* June 11, 2012, UNESCO House Room X, Paris. Print.

Trungpa, Chogyam. *Cutting Through Spiritual Materialism.* Boston: Shambala, 2002. Print.

Turkle, Sherry. *Alone Together: Why We Expect More from Technology and Less from Each Other.* New York: Basic Books, 2011. Print.

"Two Years After Andaman Tribe Dies, Another 'Faces Extinction.'" *Survival International.* n.p., Jan 26, 2012. Web. Nov 23, 2012.

"UN Global Compact Turns a Blind Eye to Corporate Malpractices." *Friends of the Earth International Website.* FOEI, May 10, 2012. Web. May 19, 2012.

DOI: 10.1057/9781137308993

United Nations. Ad Hoc Committee on an International Convention against the Reproductive Cloning of Human Beings. *United Nations declaration on Human Cloning.* Mar 8, 2005. Web. Nov 28, 2012.

_____. Commission on Environment and Development. *Our Common Future: Report of the World Commission on Environment and Development.* New York: Oxford UP, 1987. Print.

_____. Department of Economic and Social Affairs, Population Division. *World Population to 2300.* New York: United Nations, 2004. Print.

_____. "Disarmament Resource Guide: Global Issues." *United Nations Family of Sites.* United Nations Office at Geneva, n.d. Web. Nov 28, 2012.

_____. "The Feminization of Poverty." *United Nations Family of Sites.* United Nations Department of Public Information, May 2011. Web.

_____. "Global Warming Potentials." *UNFCCC Website.* United Nations Framework Convention on Climate Change, n.d. Web. May 16, 2012.

_____. "International Human Development Indicators." *Human Development Reports.* United Nations Development Programme, 2011. Web. Sep 8, 2012.

_____. "Keeping the Promise: A Forward-Looking Review to Promote an Agreed Action Agenda to Achieve the Millennium Development Goals by 2015." *Report of the Secretary-General* (A/64/665), n.d. Web. Nov 27, 2012.

_____. "Measuring inequality: Gender-related Development Index (GDI) and Gender Empowerment Measure (GEM)." *Human Development Reports.* United Nations Development Programme, n.d. Web. Nov 28, 2012.

_____. "Millennium Developmental Goals: A Gateway to the UN System's Work on the MDGs." *United Nations Family of Sites.* United Nations, 2010. Web. Nov 28, 2012.

_____. "Multidimensional Poverty Index (MPI)." *Human Development Reports.* United Nations, 2011. Web. Nov 28, 2012.

_____. *Resilient People, Resilient Planet: A Future Worth Choosing: The Report of the United Nations Secretary-General's High-Level Panel on Global Sustainability.* New York: United Nations, 2012. Print.

_____. Sixty-Fifth General Assembly. "Sustainable Development." September 2010.

_____. Sixty-Sixth General Assembly First Committee 18th Meeting. "If Conference on Disarmament Cannot Break 15-Year-Long Deadlock, United Nations 'Could Lose Legitimacy in Disarmament Affairs'

DOI: 10.1057/9781137308993

First Committee Told." *General Assembly GA/DIS/3443.* Oct 21, 2011. Web. Nov 28, 2012.

———. "Sustainable Development: Background." *UN General Assembly Website.* UN, n.d. Web. Nov 28, 2012.

"United Nations Indicted for Enabling Corporate Control of Water and Greenwashing." *Polaris Institute Website.* Polaris Institute, Jan 27, 2010. Web. May 21, 2010.

"Universal Declaration on Bioethics and Human Rights." *UNESCO.org.* UNESCO, Oct 19, 2005. Web. Nov 23, 2012.

University of Nebraska-Lincoln. "Military Greenhouse Gas Emissions: EPA Should Recognize Environmental Impact of Protecting Foreign Oil, Researchers Urge." *ScienceDaily Online,* Jul 22, 2010. Web. May 21, 2010.

"Updated List of Free Sources on the Global Compact." *Global Compact Critics Blog.* n.p., May 17, 2011. Web. Nov 28, 2012.

"US Military Energy Consumption." *Energy Bulletin.* n.p., May 20, 2007. Web.

Vartabedian, Ralph. "MIT Prof. Robert Solow Wins Nobel Prize in Economics." *Los Angeles Times Online.* Oct 22, 1987: n.p. Web. Nov 28, 2012.

Verhoeven, Martin J. "Buddhism and Science: Probing the Boundaries of Faith and Reason." *Religion East & West* 1 (2001): 77–97. Print.

Vinge, Vernor. "The Coming Technological Singularity: How to Survive in the Post-Human Era." *VISION-21 Symposium March 30–31, 1993.* NASA Lewis Research Center and the Ohio Aerospace Institute, 1993. Print.

Wachtel, Paul L. *The Poverty of Affluence: A Psychological Portrait of the American Way of Life.* Philadelphia: New Society, 1989. Print.

Walker, Alice. *We Are the Ones We Have Been Waiting For.* New York: The New Press, 2006. Print.

Ware, Alyn. "Nuclear Stockpiles." *NuclearFiles.org.* Nuclear Age Peace Foundation, n.d. Web. Nov 27, 2012.

West, Larry. "Fire and Ice: Melting Glaciers Trigger Earthquakes, Tsunamis and Volcanoes." *About.com: Environmental Issues.* About.com, n.d. Web.

"What Is Biomimicry?" *AskNature.* The Biomimicry 3.8 Institute, n.d. Web. Nov 23, 2012.

"What Is IPM?" North Dakota State U, May 2, 2001. Web. Nov 23, 2012.

"What Is Sustainability?" *MIT Sloan Management Review.* MIT, n.d. Web. Nov 27, 2012.

DOI: 10.1057/9781137308993

Wilcox, Richard. "United States Militarism, Global Insecurity and Environmental Destruction." n.p., Dec 2003. Web. May 21, 2012.

Wilson, Edward O. *The Social Conquest of Earth*. New York: Norton, 2012. Print.

Wincher, Tim. "UN Wants to Use Drones in DR Congo Conflict." *Agence France Presse* Nov 23, 2012. Web. Nov 28, 2012.

Wisman, Jon D., and Barton Baker. "Increasing Inequality, Status Insecurity, Ideology, and the Financial Crisis of 2008." Working paper. American University, 2009. Print.

The World Bank. "Press Release: World Bank Sees Progress Against Extreme Poverty, But Flags Vulnerabilities." *The World Bank Website*. The World Bank Group, Feb 29, 2012. Web. Nov 28, 2012.

_____. Poverty Analysis, Poverty Reduction and Equity, Web.

World Development Movement. *The End Game in Durban? How Developed Countries Bullied and Bribed to Try to Kill Kyoto*. London: World Development Movement, 2011. Print.

World Development Movement Website. World Development Movement, n.d. Web. Nov 28, 2012.

"World/Global Inequality." *Inequality.org*. Program on Inequality and the Common Good, n.d. Web. Nov 12, 2012.

World Health Organization, ' Investigation and Evaluation of Chronic Kidney Disease of Uncertain Aetiology in Sri Lanka, Final Report' (unpublished doc: RD/DOC/GC/06) 2012. Print.

"World Water Crisis: >1 out of 6 People Lack Safe Drinking Water, 2/3 of World Population to Suffer for Water Shortages by 2025." *WcP Blog: World Culture Pictorial*. n.p., Jul 16, 2011. Web.

Zabarenko, Deborah. *Countries Must Plan for Climate Refugees: Report*. Reuters: Thomson Reuters, Oct 27, 2011. Web.

Zajonc, Arthur. "Cognitive-Affective Connections in Teaching and Learning: The Relationship Between Love and Knowledge." *Journal of Cognitive Affective Learning* 3.1 (2006): 1–9. Print.

DOI: 10.1057/9781137308993

Index

DOI: 10.1057/9781137308993

DOI: 10.1057/9781137308993

DOI: 10.1057/9781137308993

DOI: 10.1057/9781137308993

DOI: 10.1057/9781137308993

DOI: 10.1057/9781137308993